How Typography Works

How Typography Works

(and why it is important)

Fernand Baudin

Foreword by Charles Bigelow
Department of Computer Science
Stanford University

Design Press

First U.S. Edition, First Printing

Copyright © 1984 Editions Retz, Paris
Copyright © 1988 Lund Humphries Publishers Ltd,
London
English translation © Fernand Baudin

Printed in England

Design Press offers posters and The Cropper, a device
for cropping artwork, for sale. For information, contact
Mail-order Department. Design Press books are available
at special discounts for bulk purchases for sales
promotions, fund raisers, or premiums. For details contact
Special Sales manager. Questions regarding the content
of this book should be addressed to:

Design Press
Division of TAB BOOKS Inc.
10 East 21st Street, 1101
New York, NY 10010

Library of Congress Cataloging-in-Publication Data

Baudin, Fernand.
    [Typographie au tableau noir. English]
    How typography works : and why it is important / Fernand Baudin :
foreword by Charles Bigelow.
       p. cm.
    Translation of: La typographie au tableau noir.
    Bibliography: p
    ISBN 0-8306-5011-3
    1. Printing, Practical—Layout.   2. Type and type-founding.
I. Title.
Z246.B3813   1989
682.2′252—dc19   89-31131
                          CIP

Contents

In memoriam
Charles Peignot
& Maximilien Vox

and as a tribute
to all typedesigners

Foreword

Foreword. *How Typography Works* is an English-language version of *La Typographie au Tableau Noir* ('Typography on the Blackboard') which was first published in French by Retz in 1984. I am pleased that an English-language publisher has recognized the merits of Fernand Baudin's approach to what might be termed 'vernacular typography', and I am happy to contribute, by way of a foreword, the comments I first set out in my review of the French edition, which appeared in *Fine Print*, Volume 11, No. 1 in January 1985. I am grateful to the editor of that journal for permission to reproduce my review article more or less word for word.

Five years ago, the Belgian typographer Fernand Baudin spent a week at Stanford University as a visiting lecturer in the departments of art and computer science, where he spoke on the subject of 'Constellations of Typefaces and Configurations of Text'. This was the first time that most of his English-speaking American audience had heard these ideas, though he had earlier contributed an article on the same subject to a sumptuously produced, limited edition typographic anthology, *De Plomb, d'Encre, et de Lumière*, published by the French Imprimerie Nationale in 1982.

Baudin's thesis in his Stanford talks and in his article is at once simple and profound. He claims that literacy involves not only a knowledge of how to form letters and assemble them into strings, but also an understanding of how to organize the elements of the text into a coherent visual structure. This latter task he calls the 'visual editing of text'. It follows that typographic design, layout, mise-en-page, etc., are simply aspects of the visual editing of text, and that the principles behind what often appear to be the arcane and specialized practices of professional typographers are in fact the common province of every literate person.

These are exciting, heartening, liberating, and sobering ideas. Exciting, because at one stroke handwriting, calligraphy and typography are subsumed in a grander view of literacy that preserves their differences while emphasizing their similarities. Heartening, because in a modern world that first replaced formal handwriting with the typewriter and is now replacing the typewriter with the computer printer, we need not fear that the dissemination of typography to the masses will necessarily degrade the visual quality of the text. Liberating, because we now need not dwell on the specific technology of writing, whether pen, typewriter, laser printer, or typesetter, but on the principles of the text, in order to understand the techniques of literacy. Sobering, because we must realize that education, not technology, is the prime mover of literacy, and that we have allowed our schools, through the increasing degradation of the role of the teacher, to fall behind the technical ingenuity of our computer industry.

During his Stanford visit, Baudin showed the manuscript of a book he was writing on this subject. It was intended to be, among other things, a manual for teachers as well as a general introduction to the subject of letterforms and layout. The manuscript was written in Baudin's own vigorous hand, and illustrated in most cases by his own rough sketches. That manuscript was published as the book, *La Typographie au Tableau Noir* ('Blackboard Typography'). The publisher and author deliberately chose to reproduce the handwritten manuscript by photo-offset rather than to typeset the text. In most cases, the author's original illustrations were also reproduced. The result is an extremely lively work in which the spirited personality of the author is visually evident on every page.

At one level, the book is a useful general introduction to the principles of typographic design, and could easily serve as a text or reference book for classes in typography and graphic design. It discusses the variety of letterforms and the principles of their design, stonecut inscriptions, the layout of pages from papyrus fragments to the modern newspaper, and the ideas of symmetry and asymmetry. At a deeper level, it explores Baudin's philosophical ideas on the meaning of literacy in the modern age – or, in fact, in any age. It is a treatise on 'blackboard' typography because it is quick, rough, and ephemeral, and the visual appearance of the text makes this clear. Baudin wants us to understand visually as well as conceptually.

The direct reproduction of handwriting is initially somewhat shocking to eyes accustomed to reading typographic books, and yet it has been a goal of many letterform educators and lettering artists. I remember vividly Lloyd Reynolds's passionate advocacy of photo-offset printing for calligraphy and handwriting. The good designer knows that rough work may be better and stronger than smoothly finished work, if the underlying principles are strong and correct. What Baudin is proposing is a vernacular typography, a typography in which the literate person has control of the tools of production, whether it is the pen, the typesetter, or the laser printer. This is the kind of approach we must take soon, if we want our traditional standards of typographic legibility to be transferred to the new technologies of computer printing which are taking over almost every aspect of printed communication.

Charles Bigelow
Stanford University

In all countries,
the education of the young
& everything pertaining to it:
handwriting, printing & books,
always seemed to me
an object as vital to the
interests of the Prince
its Finances or any other
consideration whatsoever.

ÇAYAS
Secretary of Philip II
to Chr. Plantin, 1567

Introduction

Introduction. At the end of this book, on pages 133–134, I explain how the volume came to be written – in the literal as well as the conceptual sense. I welcome this opportunity to offer the ideas originally put forward in the French edition to an English-speaking audience.

What exactly am I setting out to achieve? At first glance, the book offers a journey through the landscape of typography, pausing spread by spread to take a closer look at some of the underlying concepts. Professional typographers may regard this as rather trite, or 'prentice work'. However, it is not professional typographers I am addressing, but those concerned generally with education. With the advent of desk-top publishing more and more of us will at some stage become involved in the process of communication by generating passages of text on a computer or word-processor. But we are all consumers of typography, and so a grasp of the underlying principles of typography – the nature of individual characters, the assembly of characters into words, of words into sentences, and the arrangement of passages of text on a page – insofar as they assist in a proper understanding of what is being communicated, is a vital component of literacy. Therefore, more than ever before, education should help us in ordering and arranging whatever we have to discuss either orally or in writing.

It is a curious paradox that precisely at the moment when the technology of communication by way of computer-generated typesetting and computer database storage facilities is accelerating at a furious rate, there is no corresponding acceleration in the spread of literacy. It is a sad fact that literacy may actually be in decline. All the more reason, then, that those who are not professional typographers but who have a keen interest in education, and the printed word, should have an opportunity to familiarize themselves with the common currency of typography we all take for granted.

Typesetting and typewriting. The introduction of compulsory education, a century ago, happened to coincide with the invention of mechanical composition and typewriting.

In 1868, C. L. Sholes patented the typewriter as we know it, and in 1873 awarded the contract for its production to Remington. In 1886 Ottmar Mergenthaler completed development of the first Blower Linotype machine which was installed at the *New York Tribune*. In 1887 Tolbert Lanston was responsible for the launch of the competing Monotype system. The Education Act which introduced compulsory universal education in Britain passed into law in 1870. The developments in mechanical typesetting effectively brought to an end four hundred years of setting type by hand. The typewriter marked the start of the mechanization of the office and the replacement of handwriting as a means of business communication and the method by which business records were compiled and stored. It is hardly surprising that handwriting and calligraphy came to be considered as obsolete, at least temporarily. In the hundred years that have passed since these events, we have witnessed other developments. By the 1960s phototypesetting machines were beginning to supersede the hot-metal machines which were the lineal descendants of those first Mergenthaler and Lanston inventions. But whereas it took seventy years for mechanical composition to be overtaken by photocomposition, it only took another twenty five years for phototypesetting in turn to be overtaken by a generation of machines which relied for their image-formation not on photo-matrices but on digital CRT technology. This technology in turn is being supplanted by raster imagesetters using digital founts and laser exposure.

Meanwhile in the office environment, new generations of typewriter have appeared embodying first electricity and then electronics, and paving the way for desk-top publishing equipment based on computers which can in some instances approach the printers' typesetting equipment in terms of versatility. The gap between the appearance of text created by typewriter in the office and by typesetter in the printing house or typesetting firm has narrowed. But initially they were thought of as distinct processes with different uses: the skills were taught separately, too. It was clear that the typewriter was not in competition with typesetting – any more than shorthand was in competition with longhand. It was realized that typesetting and typewriting involved different modes of production to serve different reading situations. The typewriter was restricted to documents involving one typeface only, in one size, on one paper format and one side only of any given page. Typesetting on the other hand, offered a vast constellation of typefaces in a wide range of sizes, weights and formats in order to deal with the design of books of many descriptions and texts of any configuration. Moreover, the typewriter called for only one fixed margin, on the left-hand side of the page. Typesetting needed four predetermined margins per page on both sides of a leaf – with all the inherent problems of so-called imposition.

In mentioning the two technologies available for multiplying copies – typesetting and typewriting – we must not overlook a third and older 'technology' of communication: writing, in the visual as well as in the conceptual sense.

Before typesetting, before typewriting, before writing, there was of course an oral tradition of communication accompanied by pictures and pictographs. It lasted for hundreds of thousands of years. Writing has only been available to man for a mere 5,000 years. Yet it made all the difference between history and prehistory.

Writing initially was virtually the monopoly of scribes and monks working for secular and religious authorities. The ability to write and to read was strictly circumscribed. The pen was in many instances mightier than the sword, and this power was retained by the authorities of Court and Church. The preparation, and in some cases the illumination, of manuscripts by hand, the art of calligraphy, produced in some cases works of outstanding beauty and excellence, and models for others to follow. Then came Gutenberg.

Writing without a pen. For five centuries Typography (the term we use to describe the process and appearance of typesetting) has been more particularly associated with the production of books. In the thirteenth century western universities developed a system of piece-work in order to multiply handwritten copies of any given exemplar or approved version of a book. Whether or not Gutenberg was really the inventor we suppose him to have been makes no difference to the fact that what was invented was not a new alphabet or a new layout but a new method for multiplying identical copies of any constellation of movable type and any configuration of text. Henceforth type design and typesetting went their separate ways. This new method was not called Typography by contemporaries. It was called 'the art of printing and writing without a pen'. In the French of Christopher Plantin, 1567: *écrire à la presse sans plume*. The word *Typographie* to designate a printing shop was introduced in France as early as the mid-sixteenth century. Only in late seventeenth-century England did Typography begin to imply the theory and practice of type design, typecutting, typecasting, typesetting and printing. Today it tends to suggest the use of typefaces in communication generally. The invention and use of movable type marked a watershed in the history of writing and publishing because it permanently disrupted the ecology of writing as well as the economics of book production. As a result we have concentrated on the professional training of specialists and totally overlooked writing (with or without a pen) as an essential part of our heritage.

The typographic heritage. Today we have available any number of methods for writing almost anything without a pen and for transmitting the message at amazing speed. It even appears at times as if any computer-assisted keyboard

ABCDEFGHIJKLM

Baseline indication between letters

NOPQRSTUVWXYZ

& Æ Œ Ø Ç Ł

ƒ£$$1234567890°¢#%@

1234567890 aeilmorst

Superiors in color for use in fractions. etc.

§/!?«».,:;'""·⁚⁝⁞... ~ ˘ ° ˙ ()[]

abcdefghijklmnop

qrstuvwxyz ß fi

æ œ ø ç ł * † ‡

1234567890¢£$1234567890

Lining Figures slightly larger than lowercase x-height...... or......Oldstyle Ranging Figures

Small Caps in weights appropriate for text composition

ABCDEFGHIJKLMNOPQ

RSTUVWXYZ

&ÆŒØÇŁ .,:;'""!?

Minimum stroke thickness at 3½ inch Cap height ☞ ▬▬▬▬▬

A complete font from the International Typeface Corporation: upper-case, lower-case, figures, accents and punctuation marks gives direct access to the typographic heritage of five centuries. But that is only true as far as it goes. It is rather like saying that any piano gives direct access to the bulk of our musical heritage. To what extent this is true can be heard in the case of music as soon as anyone touches the keys of a piano. It can be seen in the case of typography whenever a keyboard operator is left to his or her own devices with a typographic problem to solve. Nevertheless the development of typewriting and word-processing has misled some educators into believing that the craft of handwriting has become obsolete and altogether expendable. This amounts to aiding and abetting illiteracy precisely at the moment when the need for and means of written communication are expanding at a phenomenal rate. The opposite extreme would be to suggest that in order to meet the communication challenges of the future everyone should be trained as a professional typographer. Which is nonsense, of course. What is needed in the circumstances is a wider critical knowledge and understanding of typography amongst the public, if only in order to value the heritage in the first place and to turn it to good account. And to share common standards of quality.

Culture and technology. In order to value and assess our typographic heritage we must first make an inventory. That is what this book will help you to do: to perceive how and why typefaces are different and call for more visual editing than handwriting or typewriting. At the moment what is mystifying for many people who are involved in typography as distinct from typewriting, is that they have not been told, for example, that space, i.e. spacing between characters and words, 'leading' between lines, and margins around text are just as important as the types themselves. Nor have they been told that sobriety is of the essence when confronted with such a wealth of letterforms. Or that maximum legibility is a result of orderliness, simplicity and clarity in the treatment of text. This is so, regardless of the complexity of the technology and of the subject matter. It is not the case that our typesetting technology should attempt to emulate the parchment, the gold leaf, the miniatures and the illuminations of medieval and renaissance manuscripts. It is the case that the visual efficiency i.e. the *legibility* of our written communications must match the efficiency of our typesetting technology. This is a question of visual culture being perceived as the indispensable complement of grammar and technology.

Visual literacy. A century after the introduction of compulsory education, typewriting and mechanical composing machines, there is admittedly no corresponding progress in the general level of literacy. On the contrary, illiteracy is spreading at such a rate that Daniel J. Boorstin, sometime Director of the Library of Congress, in Washington, is not the only one to consider that not only typography and book production but Western industry and the economy as a whole are in peril. This is as much as to say that the whole fabric of Western culture and civilization is in peril, since it is entirely dependent on written communication and telecommunication.

The ability to write is the initial step towards any further development of the individual as a fully-fledged citizen in a modern democracy. This is so because in a culture of the written word handwriting is, of course, the one direct access to writing as a mental process and as a practical activity. Proficiency in writing as a mental process, however, does not necessarily imply a corresponding proficiency in either the art of calligraphy or the practical science of printing. But there is a degree of literacy which implies an awareness of letterforms for what they are: namely a vital part of our daily environment and the instruments of efficiency in written communication considered as information, a decoration or both. In a democracy there is no reason why this awareness should be the preserve of any particular category of citizens any more than grammar or syntax.

Typographical literacy. The essence of writing and lettering is to make language visible and retrievable: *verba volant scripta manent*, spoken words pass away, written words are there to stay. The consequences of that process are so far-reaching that they are incalculable. It affects not only language itself, it affects every aspect of our culture and civilization.

Typography is the technology which has transmitted and augmented the finest heritage of calligraphy. That is why it is worthwhile to value typefaces as the essence of typography, as the essential tools which have survived all the technological revolutions. Typography now encompasses design with type-faces as well as the design of typefaces. It no longer implies typecutting, and typecasting. But the precision of typefaces as such still implies a corresponding precision in their visual editing, i.e. in the choice of typefaces, their spacing and their placing on the page to achieve a given purpose. In other words they imply the highest degree of precision in the visual as well as in the grammatical structure.

Education and professional training. The English version of this book as well as the original French addresses the self-taught 'typographers' among the educators as well as among the general public. It is hoped that it will help more people to understand what typography as distinct from calligraphy and typewriting is all about. A century ago it was still possible to believe that calligraphy so-called could help you to find a job. Or that a printer would take care of your 'copy' whatever it looked like. This is no longer so. Since then the technologies of writing as well as the technologies of printing have been disrupted several times over. As a result of computer science both categories of technology are accessible to more and more people less and less connected by common professional values and cultural criteria. How to connect more and more people by common professional values and cultural criteria is clearly what education is all about. In all cultures of the written word, education is synonymous with literacy. If the essence of typography is the typeface (as against the older technologies of typecutting and typecasting) then computer science which eliminated those technologies is clearly the one technology best suited to spreading typographic, i.e. cultural, values.

Fernand Baudin
1988

For all writing
is good or worthwhile—
only according to
the amount of service
one gets from it
to conduct ordinary
affairs more ably.

PLANTIN

*Dialogues françois
pour les jeunes
enfans. 1567*

How Typography Works

For grammarians & linguists the alphabet means a system of 26 phonetic symbols. For the graphic designer & the typographer, it is a system of signs with a visual logic all of its own. The number of signs may vary from 34 to 100 or more depending on the nature of the assignment. An alphabet of 26 majuscules & 8 punctuation marks will do for a signboard or banner. 26 minuscules, 26 majuscules, 10 numerals, 18 punctuation marks, 8 ligatures (ffi), 28 accented majuscules, 28 accented minuscules, in short, a FONT of 144 SORTS, may fall short of some specific tasks. In any case, you need italics in equal numbers. Not to mention special symbols in the several sciences. All this calls for a common visual logic. There is nothing mysterious in all this. It is not even complicated. It is essentially a matter of rhythm in the writing hand. Of coherence in the visual editing of the text matter: in the choice of characters & in their arrangement in space. That is what we want to make clear.

You can draw no end of vertical parallels, thick or thin. They will just look like a series of parallel lines. Forget geometry, abstraction. Give your lines colour & texture. Add some imagination & artwork: you can combine them in any number of patterns. But they will still look like vertical bars, stripes, streaks or strokes. That will never make them into signs or symbols with a reference to sounds, to sense or meaning. Except, of course, as bar codes.

It is enough to add a dot & all these vertical bars, stripes, streaks or strokes will immediately suggest the appearance & the sound of the letter i. You can repeat, the i's, alter them, modify their relative positions, their height, their colour, their thickness & once more make them into endless visual patterns. However pleasant they may be they will never suggest another sound. Another letterform. One i is not enough to form the basis for a whole alphabet of related letterforms.

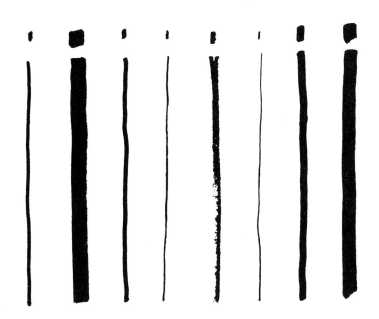

Bring together a vertical bar & a circle, & they may well suggest the letter i & the letter o. Make both of them thick or thin, condensed or expanded, and you introduce the essential notion of 'colour' in the typographic sense. Quite obviously a page covered with rows of i's & o's will differ in appearance depending on whether the characters are bold or light. These are the rudiments of what Maximilien Vox called 'visual phonetics'.

If you add the so-called SERIFS you can make a clear distinction between CAPITALS (uppercase letters) & small letters (lower-case). The origin of the serifs is historical. Their usefulness has been challenged. In vain. It is also beyond dispute that they slow down the writing process'. Their presence is evidence of a 'monumental' intention & makes a sharp contrast with the swift ligatures of the cursive. But, cursive or monumental, any letter form can be precise or slurred.

In the top row the I's are upper-case (CAPITALS). In the bottom row they are all lower-case (minuscule). It is clearly not a matter of dimensions. The difference is in the number & the direction of the serifs. You can date a letterform by its serifs. Their finish or lack of it reveals the talent of the craftsman who executes the task. Or the intentions of the user (who may well be a different person).

Upper-case or lower-case O's have no serifs.* Even so they are full circle or narrow, thick or thin, in accordance with the fullness, narrowness, thickness or thinness of the I's in the same FAMILY. The several O's & I's on this double page spread are correct all of them, but each one belongs in a different family. Anywhere else it would stick out like a sore thumb.

*
the use of the terms upper- and lower-case derives from the trays in which the compositor used to store his metal type.

NB. In isolation an O cannot be identified as upper- or lower-case. The same applies to v, w, x, z

The graphical order is not the alphabetical order. They are altogether ~~different~~. The former is determined by the similarities between characters made up of mainly straight lines, circles or obliques. The narrowest letter is always the i. The widest ones are always the M & W. When thinness is required it is necessary to accentuate the contrast between narrow & wider letterforms.

Where the E is only one half the width of the W, the proportion, the "rhythm" is good.

Where the E has the same width as the W, the proportion & the optical "rhythm" are bad.

L F E P R B S are approximately one half the width of an N or an H.

IJLFE PBRS

OQCGD

MW

HUNT

AVXYZK

IO

IJLFE BPRS MW

OQCGD HUNT AVXYZK

To win space one has to give up
contrast, rhythm & colour.
One makes do with the mono-
tony of condensed typefaces.
Only the M & W overlap the
narrow limits of the common
measure.

IJLFE PBRS

O Q C G D

HUNT MW

AVXYZK

To ensure more impact & colour in the headings, additional thickness & width are given to the individual characters (this in turn affects the style & grammar of the sub-editors).

Apart from proportion, colour, serifs, use can be made of contrast between thicks & thins. The distribution of thicks & thins is governed by age-old reading habits which it is safer to respect for the sake of legibility. Certain debasements are inevitable. This may be due to ignorance or to a striving for effect which may be more or less warranted by talent & circumstances.

good   bad

IJLFE PBRS S

OQCGD
Q

MW MMW

inverted   faulty   invert

HUNT   *correct*

HUΛN   *faulty*

AVXYZK   *correct*

AAXYZ   *faulty*

inverted   faulty   inverted   inverted   faulty

MBSEF   *faulty*

inverted W   top heavy   top heavy   faulty   faulty

The upper-case alphabet of roman capitals is descended from Antiquity. To this day it bears the stamp of the deliberate & monumental style of Imperial inscriptions on the triumphal arches. Haste is excluded, as well as ligatures. Every single letter is built up stroke by stroke. Utter care is devoted to the serifs, the contrast of thick & thin. We still use them mainly for titles; as initials; for inscriptions, banners, etc.

The alphabet of minuscules is part of the heritage of the Carolingian Renaissance (ninth century AD). Upper- & lower-case are two altogether different writing-systems. All the capitals line between two horizontal parallels. While it takes 4 horizontals to line the minuscules; two for the normal letters: a e i m n o r s v w x z; one for the ascenders: b d f h k l; one for the descenders: g j p y.

inmulhrtj

oabdceg

pqsfa g

vwxyzk

Just as with the capitals, the colour of the minuscules can be modified. By ignorance, or deliberately, to invite different connotations, to suggest different atmospheres this can be done by tampering with the serifs, with the thicks & thins, etc. Here again, conventions & habits must be respected for the sake of legibility. They cannot be ignored without good reason & a lot of compensating talent. Whatever disturbs any reading habit is detrimental to legibility & is promptly interpreted, consciously or unconsciously, as crass ignorance, hilarious blunder, witticism or aesthetic refinement.

inverted U inverted n

The ITALIC minuscule evokes the Italian Renaissance (fifteenth century A.D.) & contrasts with the
ROMAN letterforms of Antiquity. The italic & the
roman lower-case as we know them are contem-
poraneous with the beginnings of printing & follow
the same Carolingian model. The italic however is
closer to its Renaissance cursive model. Not because it has a slope, but because it is easier to
write. The Carolingian minuscule has no serifs.
It has the beginnings of ligatures. The first printers
added serifs to match the serifs of the Roman
capitals & to emphasise monumentality.

Whatever the tool, the swiftness of the movement
& the degree of slope, the first i & the first m shown
below are made in one & three strokes; whereas the other i & m imply several movements
of the hand & twists & turns of the tool to
produce the serifs. This is more elaborate, i.e. time-
consuming, i.e. more expensive. Time is money
& always was. Scribes of old counted the letters
& measured their spaces. What publishers, printers
& typographers to-day call casting-off, the Roman
scribes called stichometry. The amount of time
& money involved is never negligible.

abcdefg
abcdefg
hijklm
hijklm

nopqrstu

nopqrtu

he italic is NOT
sloped version of
te roman as can
= seen when a
mparison is made
etween individual
aracters & their
troke -formation

vwxyz

Arabic numerals are part & parcel of the Western writing-system. So much so that scientists & technicians prefer the term 'alphanumeric' to 'alphabetic'. The numerals must be redrawn from 1 to 0 for each new typeface: in roman, in italic, in bold, etc. There are two kinds: lining figures which range with capitals & non-lining or old-style figures which range with the lower-case.

1234567890

1234567890

Lining figures are used to match capitals, for example in letterheads, for some old-style figures (e.g. 1, 2, 0) could look too small. Old-style figures are used to match lower-case setting, in case lining figures appear too large.

Punctuation marks must be designed for each new typeface. Just as in the case of figures they must match the letterforms in their shapes, in their 'weight' & in their overall typographic colour.

$$ (: ; , ! ? \text{ '' ''} /\S \ll \gg)^{*} $$

Note the hyphen & the dashes of two alternative lengths. Also the inverted commas (to be used singly or in pairs) & the continental quotation marks in which German usage (» «) is the opposite of French usage (« »).

$$ \$ \; \pounds \; \% $$

These are the three more familiar symbols. They too are drawn anew for each new type. Less familiar symbols are standardised.

The number of accents & ligatures varies according to the several languages, the subject matter & the kind of typeface involved. Accents vary for grammatical & linguistic reasons. Ligatures vary for optical reasons. Even in the smaller sizes the juxtaposition of two ff's, for example affects the rhythm of the composition. The curves at the top create a gap involving additional space at the foot of the second f. That is why the more frequent groupings a treated as units in their own typographic right. What is tolerable as typewritten copy is not to be tolerated as photocomposition. As soon as you have to foot the bill you understand the difference. In the same way every one understands that more care is attached to the composition of a literary work than to a parking ticket. The following are the more familiar ligatures:

æ œ ÆŒ ij ff fi fl ffi

Our alphanumeric writing-system involves far more optics than geometry. The manufacturer delivers the letters, the figures & some ligatures. It is for the user to allocate space: between the letters of a word (especially in upper-case). Between the words in a line. Between the lines in a column of text (leading as it is called). In the margins around an inscription. (Between & around the columns of the body text in any double-spread which constitutes the optical unit in book-design. In this the responsibility of the user is at least as important as the responsibility of the manufacturer,

These two horizontals are geometrically the same length. Yet, they look unequal because of a couple of additional obliques.

HEHO

For typographical reasons, i.e. optical reasons, vertical strokes call for more interspace than curves. In fine printing capitals (and the larger sizes of lower-case) are always interspaced, & given all the optical corrections needed to ensure an overall even 'colour'.

HANNIBAL
HANNIBAL

minou
minou

In the smaller sizes the spacing between words should never exceed the width of a lower-case e. This, however, does not apply to newspapers where the deadline must be met at all cost. Ever since Gutenberg new types have been developed & this prolifera- tion shows no sign of ceasing. Fashion rules type design & layout just as it rules in the field of other consumer goods. Next to speech, writing, whether it takes the form of handwriting or typography. is of the essence in our culture & society. To perceive & to value the subtleties & nuances in this as in any other aspect of civilised life is to enhance the quality of life & to enrich our culture

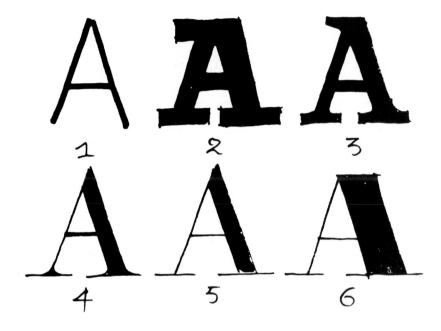

1 No serifs, no contrast

2 Slab serifs, but no contrast.

3 Serifs are robust & rounded, there is a
contrast between thick & thin.

MMMM

MMMM

4 The serifs & the summit of the A are pointed.

5 The serifs are thin & monoline; thick & thin are markedly contrasted.

6 the same, with even more contrast between thick & thin

The 'ideal' shape of every single letter is implicitly known as soon as one is able to read. The appreciation of letterforms is a matter of educating the eye which is to guide the hand. There are no rules except the essential optical corrections which become self-evident as soon as a decision has been made concerning the overall appearance of the type it is wished to design or to be used.

W W W W

W W W W

An o will always be wider than an i. M & W
are always the widest letters in any alphabet.
It is not difficult to acquire an understanding
of the modifications necessary for any letterform which has to fit into an alphabet
which is characterised as normal, bold,
light, wide, condensed, etc. These modifications
in turn call for any number of optical corrections.

ccccoe
aaaaa

All these letterforms are correct. They offer different solutions to the particular problems posed once a general decision has been made concerning the specific uses envisaged. The presence or absence of serifs, the shape of the serifs if any, the contrast & distribution of thicks & thins, the thickness of the strokes (never monoline in fact but only in appearance) all reflect a clear intention and should contribute to clear expression.

cocet
sssst

Mere contemplation will not do. Really to get the
feel of a particular kind of handwriting or letterform you must try & imitate it pen in hand.
(this simple remark of Jean Mallon caused a
sensation in palaeographic circles in the 1930's).
To understand the optics of an alphabet, better
try & reproduce it, letter by letter, in any size, but
in one & the same size for all letters of one alphabet.

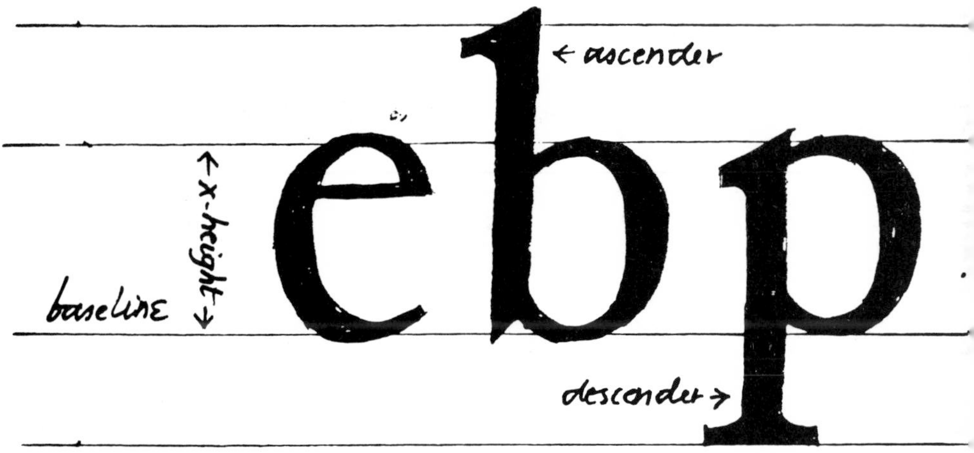

← ascender

← x-height →

baseline →

descender →

Above: letters with a small x-height. On the facing page
letters with a large x-height. The size is identical in both
cases. The difference is in the contrast between the normal
letters and the ascenders & descenders. All ascenders line
at the top. All descenders line at the bottom. But the nor
mal letters are much deeper on the facing page. And
descenders & ascenders are shorter accordingly. Tradi
tionally the smaller x-height is for book-typography. T
Italian Renaissance marked a return to the Carolingia
alphabet & the progressive elimination of Gothic letter

ebp

point size

rms. The black-letter, however, was never totally eliminated. In
many, in Switzerland it is ever-present. Nor is it absent in Britain
the u.s. It would be a real loss for typography & calligraphy if it
ere to disappear altogether. A larger x-height has been popular-
ed by newspapers. The poor quality of newsprint & the speed
f the printing process combine to fill in the bowls of a b d e
o p q . Technology may change. Not so the letterforms, the
ext columns. For more than a century the newspaper has met
he expectations of millions. It is therefore easier to carry
n than to try to change their visual habits.

A B C D E F G H I J K L M
A B C D E F G H I J K L M
N O P Q R S T U V W X Y Z
N O P Q R S T U V W X Y Z

Futura & Nobel are two sans serif typefaces without any contrast between thick & thins. The differences, however, are considerable: in the individual letterforms & in the general appearance when arranged in text blocks. The individual characters of Nobel are heavier. Since there are ± 70 signs per line and ± 30 lines per column, all the differences are multiplied by ± 2100. It is therefore no surprise that a page looks different depending on the type chosen. Different papers do not absorb ink & reflect light in the same proportions. The nature, the number & 'weight' of the illustrations if any are also decisive in the choice of the typefaces for textmatter, headings or captions.

abcdefghijklmnopqrst
abcdefghijklmnopqrst
vuwxyz 1234567890
vuwxyz 1234567890

Futura was designed by Paul Renner in 1927; Nobel, by S.H.
De Roos in 1929. Both are excellent. Yet they can look old-fashioned
or trendy depending on how they are used: at random or
to some purpose. Both were intended as book-letters i.e. for
continuous reading (discontinuous reading is associated
with dictionaries & the like). In books, today, both would look
passé, but for publicity purposes both can be very effective. In
both types, the figures line with the caps. In this connection it
is just as well to indicate that the zero figure (0) is narrower
than the letter O whether upper- or lower-case - in order that
they should never be mistaken.

What's the idea? The purpose of this book is to make a critical understanding of typography available to everyone, and more specifically to those concerned with education. The printed word used to be the preserve of well-trained professionals, but the computer has destroyed the ecology of hot-metal printing. As a result we are facing a technological explosion in written communications coupled with a growing shortage of competent communicators. This is a matter of concern for those involved in education as well as for type-designers and type-manufacturers. For five hundred years the latter have been aware that the composition of a text, i.e. the assembly of ready-made typefaces, is not a task that is natural and self-evident any more than are handwriting, typewriting or the process of type-manufacture. The skill has to be taught. It has to be learned. Moreover, typefaces come in a far more sophisticated range of shapes and sizes, weights and widths than the letterforms used for typewriter keyboards. There is another principal difficulty: type-designers and manufacturers are responsible for only one half of the final product, the 'black' forms of the text. The other half is the responsibility of the user, for it is he who decides where the 'white' spaces appear. The user 'visually edits' the individual letters, signs, words, lines and columns and decides how to place these elements on the page. Control of the position and appearance of letterforms – whether these are created as handwriting, typewriting, or typesetting – calls for fluency in the use of visual grammar. There is no longer a case for handwriting, typewriting and typesetting to be taught as separate skills: regardless of their future careers, young people should be made aware of the visual resources and possibilities of letterforms available to them. Just as they are all exposed to the same

Left: Futura. Right: Gill. Same measure (28 pica ems), same size (10pt), same leading (2pt). The difference is in the 'colour' and width. A choice between these two faces may be influenced by any number of considerations. A light illustration calls for a lighter typeface. A face that occupies more space means more paper, and paper is expensive. Hard paper is hard on the contours of the letterforms and requires more ink. Legibility is not enough. Readability is of the essence. But choice is not something to be avoided: it is positively life-enchancing.

What's the idea? The purpose of this book is to make a critical understanding of typography available to everyone, and more specifically to those concerned with education. The printed word used to be the preserve of well-trained professionals, but the computer has destroyed the ecology of hot-metal printing. As a result we are facing a technological explosion in written communications coupled with a growing shortage of competent communicators. This is a matter of concern for those involved in education as well as for type-designers and type-manufacturers. For five hundred years the latter have been aware that the composition of a text, i.e. the assembly of ready-made typefaces, is not a task that is natural and self-evident any more than are handwriting, typewriting or the process of type-manufacture. The skill has to be taught. It has to be learned. Moreover, typefaces come in a far more sophisticated range of shapes and sizes, weights and widths than the letterforms used for typewriter keyboards. There is another principal difficulty: type-designers and manufacturers are responsible for only one half of the final product, the 'black' forms of the text. The other half is the responsibility of the user, for it is he who decides where the 'white' spaces appear. The user 'visually edits' the individual letters, signs, words, lines and columns and decides how to place these elements on the page. Control of the position and appearance of letterforms – whether these are created as handwriting, typewriting, or typesetting – calls for fluency in the use of visual grammar. There is no longer a case for handwriting, typewriting and typesetting to be taught as separate skills: regardless of their future careers, young people should be made aware of the visual resources and possibilities of letterforms available to them. Just as they are all exposed to the same grammar regardless of the fact that not all of them intend to become actors, lawyers or poets. We know next to nothing about the inventor or inventors of writing as such. But we know almost everything about the inventors of the letterforms and typefaces we read and write every day. Far from being wretched scribblers, they were poets, humanists and princes in every sense of the phrase. Petrarch and Boccaccio have pride of place. They are followed by a series of popes, cardinals or chancellors such as Niccolo Niccoli and Poggio Bracciolini. Not forgetting patrons such as the Medicis. Guelphs and Ghibellines alike reformed their personal handwritings as well as the styles of their chanceries. To them this was just as important as a new style in architecture. Judging from the results they did well on both counts – and some more. But while we no longer live or build as in the days of Alberti, Brunelleschi or Bramante, we still very much use their

SOIR

We have already mentioned the importance of space: spacing, leading & margins which are all the responsibility of the user. There is another concept too: counterpunches. This technical term taken from punch-cutting denotes the hollow parts in the punches which were used in the days of hot metal to strike the individual characters in any number of matrices. The same word is still currently used to denote the white spaces inside the bowls of a b c d e g o p q. They are to letterforms what the right side & the wrong side are to a piece of textile material: they are inseparable. There is nothing the user can do about them. But he must be aware of the fact that the contours of the counterpunches are just as important as the contours of the letter

roper. Not only for type-designers but for readers generally. Indeed,
ll the care attached to the lay out & the miscription of any kind
f text relates essentially to matching the fixed spaces of the
ounterpunches & the random spaces between the letters & the
vords on any given surface: paper, parchment, stone, wood or
hatever. So it was from the earliest forms of writing, is now
- ever shall be, regardless of the technology involved. This comes
aturally whenever the scribe or miscriber & the designer are one
& the same person. The difficulties begin where there is a division
of labour. It is in this context that a common education in letterforms & typographic communication becomes as important as a
common education in the grammar & rhetoric of spoken language.

Gloria

Any alphabet is designed to some purpose, to meet a particular
need i.e. a specific reading situation. Type-design is a highly
specialised art & craft. The clients of old were royalty & princes
of the Church. Nowadays they are manufacturers of typesetting equipment. Or publishers, more or less specialised
The reading situations can & may be reduced to three
main categories: global (billboards, signs, inscriptions...)
discontinuous (directories, dictionaries, newspapers.)
continuous (novels, textbooks...)
Apart from their functional uses, letterforms & their layout take on symbolic overtones mostly unexpected & un-
predictable. Generally because they are more or less conscious

ut closely related to particular circumstances which have left
a permanent association in the minds of men & women.
hus black letter has been for centuries so closely related to
omp & circumstance that even today we go back to this letterform whenever we want to lend status & authority to a text.
Conversely & in an altogether different vein the computer has
made such an impact on the collective mind that letter-
orms designed for optical character recognition (OCR) are used
n the covers of science-fiction books. However no-one will ever
use black letter to list his available stock of whatever. Nor will
anyone in his senses use OCR type to print a missal. There are
uch notions as fitness for purpose, congeniality, acceptability.

heChip

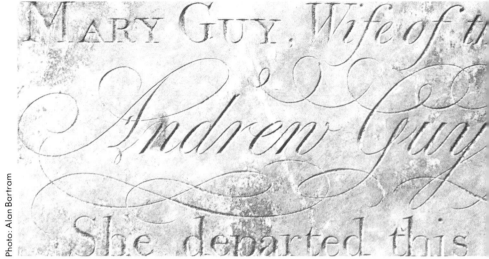

The simple legibility of an inscription reflects the attitudes of a community as well as the lack of skill of the stone cutter. The three main reading situations do not exclude any number & degrees of legibility & acceptability. Nor can they, any more than matters of taste or discrimination, be considered in isolation. Furthermore there is the legibility of a newspaper as distinct from that of a book, a cable or a neon sign. A quick note, a mere scribble, may be thrown in the dustbin or it may be typed, filed & eventually used in a photostatted memo or in book form. There are countless nuances in visual editing. They must be cultivated. They cannot be measured or listed

A label, a roadsign, a poster, a sign system can be anything between the adhoc impromptu work of a schoolboy or a school girl & the sophisticated production of an art studio. They may begin by mobilising teams of highly specialised craftsmen & end up by mobilising & directing armies & fleets. The clarity & efficiency of a sign system also manifest the intelligence & skill of the individual, the team, the nation who sponsor & supervise the project. In typography just as in writing, clarity is NEVER the result of a random happening. It calls for a combination of inborn talent & hard won experience.

Crosby, Fletcher, Forbes, *A sign systems manual*, Studio Vista, 1970

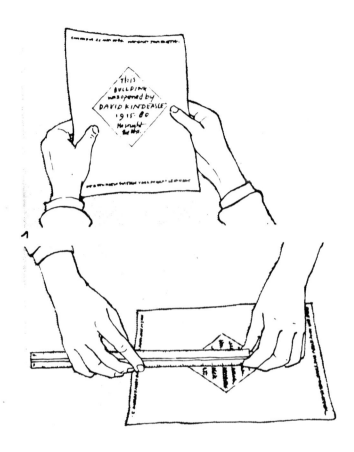

1

2

A client will find a talent to his liking.
Both may be content with a mere
scale drawing. The following sequence
illustrates the ensuing procedure (1,2).

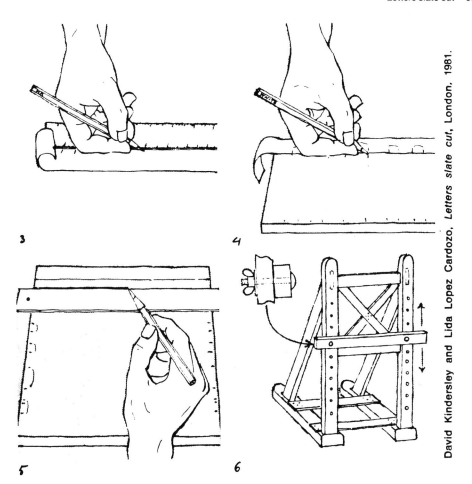

David Kindersley and Lida Lopez Cardozo, *Letters slate cut*, London, 1981.

3

4

5

6

A strip of paper helps transfer onto the slate the distances
between the lines (3, 4, 5). The slate is then transferred
to a strong easel in order to be cut (6);

(7) Drawing out the letters with a white crayon. (8) Selecting a chisel of the correct size. These are only a selection from 80 figs in *LETTERS SLATE CUT*. by David Kindersley & Lida Lopes Cardozo. Lund Humphries, London 1981

abcdefghijklmnop
qrstuvwxyz& Qæ
æ cf ffi g 123456789

Les familles JACOBS, GLIBERT, LINARD et JOSSART,
ont la tristesse de vous faire part du décès de

TANTE ODILE
née JACOBS
veuve de Triphon GLIBERT
survenu le 18 février 1983, dans sa 103ᵉ année.

Le service religieux, suivi de l'inhumation au cimetière du Centre, aura lieu en l'église Saint-Etienne, à Braine-l'Alleud, le **lundi 21 février 1983, à 10 heures.**

Réunion à l'église

LE PRESENT AVIS TIENT LIEU DE FAIRE-PART.

Matagne-Jossart, rue Wayez, 34
1420 Braine-l'Alleud.   406679 212

Slate-cut, wired or teleprinted, classified advertisement or obituary, every letter in an inscription is paid as piece-work. This furthers sobriety & brevity – the well-known source of wit. See also p.125.

ohn Sparrow, *Line upon line, an epigraphical anthology*, Brooke Crutchley, Cambridge University Press 1967

APRÈS AVOIR VU PÉRIR TOUTE SA FAMILLE
SON PÈRE SA MÈRE SES DEUX FRÈRES ET SA SŒUR
PAULINE DE MONTMORIN
CONSUMÉE D'UNE MALADIE DE LANGUEUR
EST VENUE MOURIR SUR CETTE TERRE ÉTRANGÈRE

F—A DE CHATEAUBRIAND A ÉLEVÉ CE MONUMENT
À SA MÉMOIRE

denoël

TITRE :

---

1   comme la craie dont se sert de nos jours le tailleur.

2   Lorsque le traçage des lignes avait été effectué après la

3   préparation de la peau, le scribe s'installait devant son

4   écritoire et taillait sa plume, faite le plus souvent d'une

5   penne d'oie provenant de l'aile gauche du volatile. La légère

6   courbure du tuyau des rémiges gauches en rend la tenue plus

7   aisée entre les doigts de la main droite, mais on peut éga-

8   lement utiliser n'importe quelle plume de la taille qui

9   convient. Le tuyau/était d'abord séché ou durci avant d'être
                          de corne/

10   fendu et taillé à la largeur d'empâtement exigée par le

Wherever typographic communication is taken seriously, pro-
fessionally calibrated typing paper is used to type copy to
measure. That is where visual editing begins. As a matter
of fact, for better or for worse, visual & grammatical editing
are inseparable. This is also where typography differs essentially from calligraphy. Calligraphy implies an individual-
istic approach. Typography implies shared standards of
excellence, common criteria & agreed conventions to
guide the necessary choice of a constellation of typefaces & configurations of text matter.

the typerule is a modern tool for an age-old practice.
Right from the beginning everything was created
according to measures & to numbers. Guesswork is
the hallmark of amateurism. the type-rule is there to
specify the type-size, the length of line (measure), the
number of lines per column, the amount of leading
(in any size from 8 to 14 pt) & the proportions of the
margins. This is easily done yet it is indispensable
in order to give shape & texture to the text. The latter,
not the illustrations, is the basis of the grid.

Lamentations con-
cerning the devas-
tation of Ur, Sumer,
second century B.C.
436 lines in 3 col.
Only our daily news-
papers & trade
directories use more
columns per page.
In fact these Su-
merian stereotypes
differ from what-
ever we re-hash
nowadays as
much as clay
differs from paper.

A format is either standardised or is not. Either way it dictates the choice of the materials (paper or whatever) of typefaces, typesizes & type area, etc. If it is to be a one-time inscription nothing must be taken for granted. As soon as repetition is involved, let alone mass production, standardisation is essential. It can be said with some over-simplification & saving exceptions that a book, a letterhead suggests a vertical format & long lines, while book-keeping invites oblong formats & lots of columns.

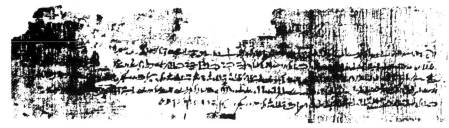

Naissance de l'écriture, exhibition catalogue, Grand Palais, Paris 1982

Deed recording an adoption. Thebes, 536 BC. Demotic, i.e. everyday script on papyrus. The long lines foreshadow the medieval charters. A vertical format on the other hand is associated with accountancy, agendas registers & the like

LE MONDE still spells its name in a fancy black-letter 45 years after THE TIMES modernised its whole appearance from the front page down to its smallest rubric. For more than a century newspapers have come in the same folio format as the more ponderous medieval books. Even so they are representative of our industrial era. (Just as the papyrus scroll represents ancient Egypt & the baked clay cylinder evokes Assyrio-Babylonian antiquities). All the news, all the styles, all kinds of type are jammed together. Space is at a premium. Forests are felled to produce the newsprint for one edition of one newspaper. Just as flocks were slaughtered to provide the skins for the manuscript of one medieval Bible. Over the past century each new technological development capable of accelerating the transmission of information has quickly been taken up: telephone, telegraph, telex, etc. This is also where the complexities of our mass culture are best reflected in the multifarious constellations of typefaces & in the varying configurations of textmatter.

Le Monde

Fondateur : Hubert Beuve-Méry   Directeur : Jacques Fauvet

3 F

M. HERNU DISSOUT
LA SÉCURITÉ MILITAIRE

Une direction chargée de missions
plus limitées sera créée

LIRE PAGE 34

Crise entre Rabat et Nouakchott

Par un sondage, et par se
...

Le sommet de Cancun

M. François Mitterrand va plaider pour des « négociations globales » sur un nouvel ordre économique

De notre envoyé spécial

Cancun. — Avant même le déb...

F. F.

(Lire la suite page 6.)

Les nationalisations

M. Jospin minimise la portée d'une éventuelle opposition du Conseil constitutionnel

JEAN-MICHEL QUATREPOINT

(Lire la suite page M.)

La fin du grand combat de M. Moussa

AU JOUR LE JOUR

Parapluie

BRUNO FRAPPAT

Le congrès socialiste de Valence

Cohérence et fidélité
par PAUL OUILÈS

Poursuivre la mutation
par ROLAND DUMAS

(Lire la suite page 2.)

HEINRICH BÖLL

Protection encombrante

ROMAN

traduit de l'allemand par
René Dadier

AUX ÉDITIONS DU SEUIL

MAGISTRATS
AUX MAINS NUES

DES AMÉRICAINS
METTENT EN ÉVIDENCE
DES GÈNES IMPLIQUÉS
DANS LE PROCESSUS CANCÉREUX

LEIRIS, MUSIL, SCIASCIA, WOOLF

Le journal intime et la création

(1) caps → medium 12pt
(2) lining → figures + 12 pt lower case
(3) lower-case medium 8pt →

5 ÉDITIONS
15 Francs
30 Pages

●●● Matin première
●● Matin
● Matin dernière
★★ Midi
★ Bourse dernière

• •

Administration	02.217 77 50	Luxembourg 16 F
Rédaction	02.217 74 80	Espagne. 70,00 Ptas
Vente et abonnements	02.217 77 50	Canaries. 80,00 Ptas
Annonces - Publicité ..	02.217 77 50	France...... 3,20 FF
Annonces téléphonées	02.217 63 29	Italie..... 900,00 L

Autres renseignements administratifs en fin de journal

o o o o

(4) lower-case bold → 18pt

France : une campagne qui vole bas

(5) light → caps 9pt

(6) lower-case bold → 8pt

DE NOTRE ENVOYÉ SPÉCIAL PERMANENT

Paris, 15 février.

« Vous avez quarante fonctionnaires en prison », lance l'un. « Rappelez-vous le S.A.C. et le massacre d'Auriol », rétorque l'autre. « Démagogue », crie celui-ci. « Menteur, boulimique », hurle celui-là. Et allez donc. A trois semaines du scrutin municipal français, la « guerre civile » fait à nouveau rage. Et les arguments échangés volent plutôt bas...

(7) lower-case roman → 7pt

L'opposition dé droite était partie, bille en tête, très fort et

Yasser Araf

Yasser Arafat à l'écoute au Con l'O.L.P., Ahmed Yamani, Yas américains, le C.N.P. ne devrait

the number, the selection & the dimensions of typefaces, titles & textmatter; the number, the selection & pro/portions of illustrations, these provide the colours of the newspaper palette. Even the smallest component in each edition is located within a graded grid, page by page. The front page is the showcase of a newspaper. It also reflects its management, its newsdesk, its readership. LE MONDE is not intended for the same readership as FRANCE-DIMANCHE. The latter is worlds apart from the FRANKFURTER ALGEMEINE, the ne plus ultra of typographic asceticism. The restyling of the LONDON TIMES by Stanley Morison in 1932 is a milestone in printing history. If only because TIMES NEW ROMAN, which was designed on that occasion under the supervision of Morison, is by now the most widely used typeface in the world. It is also the most widely copied. Copies vary as much as, and maybe more than, musical interpretations of the same concerto.

6 col. Titles in black letter and roman, centred. No big type. No photographs. No publicity. 5 articles. 7 brief news stories. No "to be continued". The ne plus ultra of sobriety. A total absence of sensationalism.

A weekly with the size of a daily & the style of a tabloid. Big titles. Big pictures. No clear grid. The rock'n roll of sensationalism as against the plain chant of contemplation.

8 col. Only one type face for all
titles & text matter. In roman
& bold. All titles are centred.
Since 1932 several restylings
have occurred, but they
have all preserved the original combination of vigour
and sobriety.

7 col. on the front page, 8 col.
inside. Two type styles for
every title & subtitle. A
wide variety of typefaces,
of justifications. Lots of
rules & boxes. Large photographs. Large advertise-
ments.

The wording, the style, the layout of a classified advertisement section obeys strict rules: as strict as the sonnet or the alexandrine. The 6-8 column grid is the characteristic configuration of the newspaper. The rubrics have a topo/graphy as well as a typography i.e. a distinctive constellation of typefaces & typesizes. A close examination of all the components in a daily newspaper amounts to a summary of the history & resources of all letterforms in their typographic versions; of all idioms of visual rhetoric. They are the same regardless of the political & philosophical opinions expressed.

Textmatter is set in house. The advertising matter is prepared outside. Initially an ad is a mere sketch, a rough. But each component is worked out by a highly qualified specialist. They all come together on the final montage. & an individual film is made for every paper where it is to appear. In our culture, this is where the best talents & the latest technology are applied daily.

THEATRES

Les Misérables
THE MUSICAL SENSATION
PALACE THEATRE
C.C. OPEN TODAY 01-379 4444 & 01-240 7200

Ukiyoe
IMAGES OF UNKNOWN JAPAN

Until 14 August 1988
Mon-Sat 10-4.50, Sun 2.30-5.50
Last admission 4.30 (5.30 on Sun)

Admission £2 (concessions £1)

For recorded information
telephone 01 580 1788

**BRITISH
MUSEUM**
Great Russell Street,
London WC1B 3DG

Every single edition of a daily newspaper is a time trial.
Clocks are ever present. There's an iron discipline. It's part
of the job. The configuration of the rubrics, the constellation
& hierarchy of typefaces & type sizes have been decided once
& for all by a graphic designer in close collaboration with the
management, the journalists & the production department.
The articles must be written to measure & delivered punc-
tually. Each article bears a distinctive mark for every
stage in its progress. At every stage handwritten corrections & specifications are coded in. The successive
proofs are measured, corrected & reprinted in order to fit
into their appointed places in the layout of the page.
All this within a span of just two hours. This description
is generally valid for most papers. With variations in
the number of editions & in equipment. The following procedure is followed in Brussels for the Belgian
daily newspaper LE SOIR.

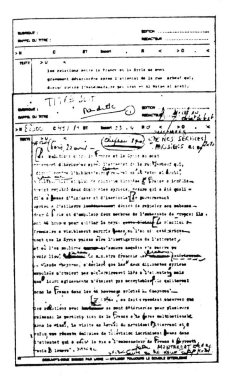

(1) Captions are typed in blue. (2) Corrections are made by hand. A subeditor gives it a number & specifies the type size, the measure, the location in the page.

(3) Articles are typed in black. Corrections are made by hand. A subeditor gives all the necessary instructions for the computer-assisted composition to be typed on a different keyboard.

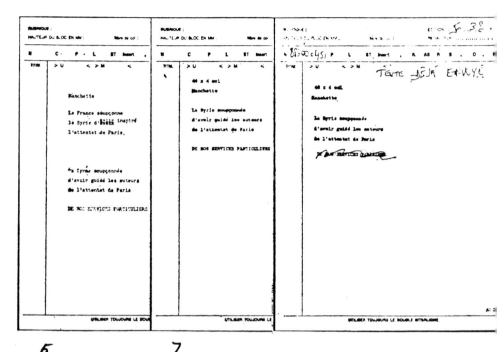

6   7

Special calibrated typing paper is printed in green for titles. They come in two versions (6). The editor makes his choice (7). A subeditor adds the handwritten instructions for the computer-assisted setting.

NB. As they mark the copies the subeditors also allocate them a place in a miniature layout (300×205 mm). The corrected films are then mounted real size (600×410mm).

9

) the output of a caption, a title & an article. The copy (9) is now interspersed with code instructions. They are intended for a keyboard where they will be followed automatically. All this in such a paradoxical way that, in order to go faster, in the end three versions are typed for every single word in any piece of copy: typewritten, computerised, typographical.

The office. Every page in every edition is given a number, a set schedule, & displayed on the wall. At any time, eight subeditors & nine make up hands can see at a glance what is the stage any given page has reached in the production line.

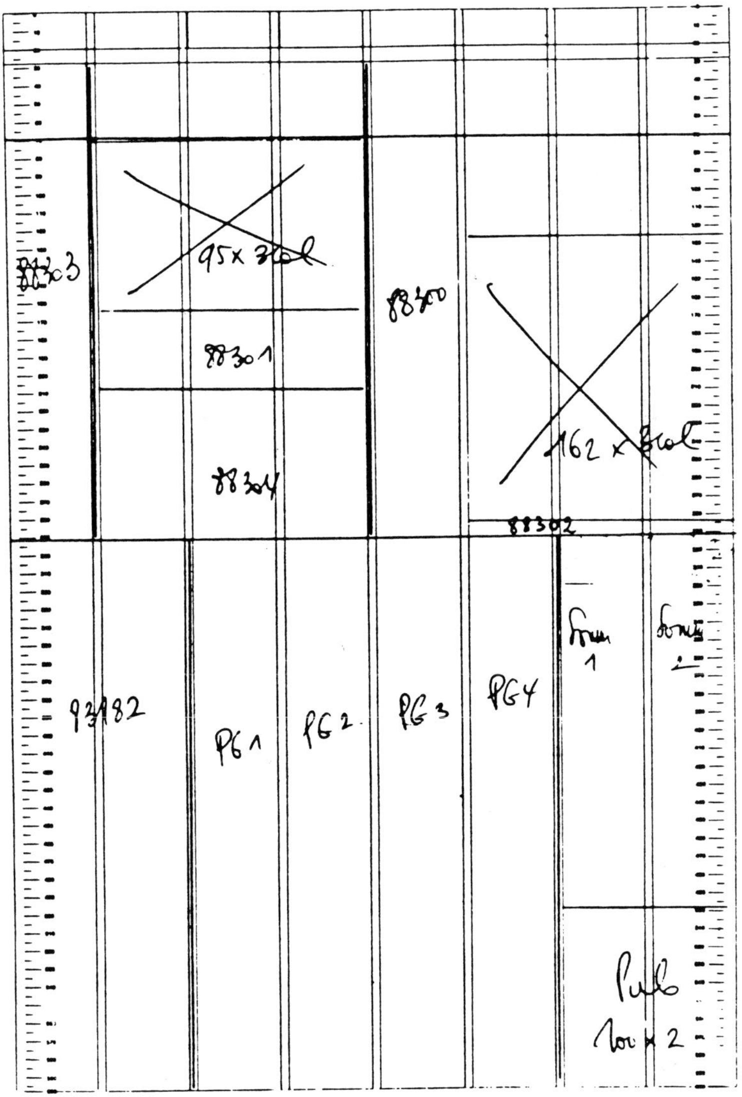
95 x 3 col
88301
88304
162 x 3 col
88302
PG 1
PG 2
PG 3
PG 4

the composing room. 54 compositors work in shifts
to key the five daily editions. This is where the
copy is coded as illustrated in fig. 7. 8. 9. On the
facing page a full scale synopsis is to be seen
where the numbers of the pages are displayed in
colour; white indicates the initial stages; red
means the plate is ready to be engraved; green
means it is press-ready.

'I this is ages away from the medieval "scriptorium". Any-
e can see that. Except for historians, however, no one pays
ny attention to the fact that here, as in the scriptorium,
am work is of the essence, as well as strict distribution of
ork along the production line. There is little room for the
nystical folklore of monk-like dilettantes or for the in-
structible reporters who manage to survive unscathed
air-raising adventures & daily cliff-hangers.

The electronic eye of this scanner reads the illus-trations which are simultaneously en-graved by a laser-beam.

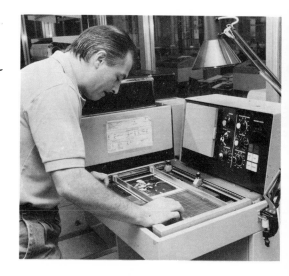

The editing terminal. The operator introdu-ces the codes which will duly allocate all the individual components in any given page.

This machine develops the first comprehensive proof of a full page.

NB. We acknowledge the assistance of LE SOIR in providing the information on which THE COPY & ITS PROGRESS is based.

This optical reader processes two full pages at a time & delivers them press-ready as polymer plates to be fastened on the cylinders of the rotary press.

Note. Since 1984, when the original French edition of this book appeared, *Le Soir*, in common with many other newspapers, has replaced the preparation of copy as typewritten pages by copy preparation on VDU screens. But the stages through which the copy needs to pass remain as described here.

(1)   (2)   (3)

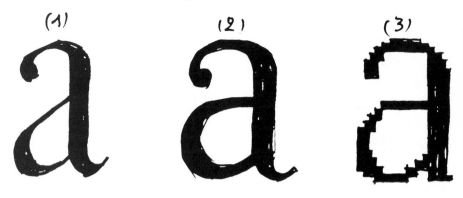

The first a recalls the sixteenth century, the golden era of typographic craftmanship. In those days only a few dozen leaves
could be printed per hour on a hand press. Ink was thick. Pape
was a handmade product. The second a is designed to be
printed with liquid ink on newsprint & to be rushed full
speed through the rotary presses. The third a was digitised
in order to be computer set. A reduction of 100 to 2 or 3 mm
smooths out the contours & reconstitutes the traditional
form. All this is not a question of 'progress' or 'decay' bu
a question of adjusting traditional forms & modern tec
nologies to meet new needs. Nor is it a question of choice.
Written communication is a vital necessity. Technology how
ever is not enough. Shared values are at least as importan

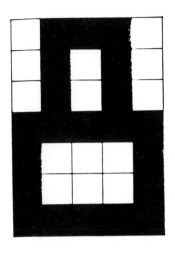

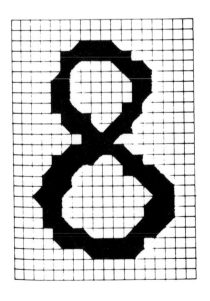

The figure 8 is all curves. The computer however ignores curves altogether & squares any circle in sight. When the definition is low the result is (1). Somewhat finer it becomes (2). And so on. For the time being economics prevail over optics. When the computer market becomes even more developed than it is, quality & cultural values will be a commercial asset again. This is a matter of years. So it was with the typewriter after 1880. For photocomposition after 1950. For computer assisted composition after 1960.

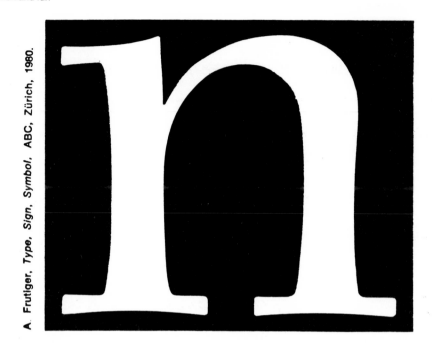

A. Frutiger, Type, Sign, Symbol, ABC, Zürich, 1980.

The letter n of the iriDiuM alphabet designed by Adrian Frutiger for photocomposition, offset & gravure. It was drawn free hand. Tools would have been a hindrance not a help. The human eye will read whatever is clear, familiar, conventional. We can & do read all the historical variants of three writing systems — & arabic numerals as well. That was too much for some talented doctrinaires who wanted to rationalise all this. In vain. Habits grow on people, become second nature, & resist mere rationality. Letterforms have at all times met innumerable social functions where legibility prevails. They also lend them—selves to the most fantastic sense of play, where legibility is hardly a consideration. There is no contradiction in this but a complementarity at all levels of individual & collective culture.

Imre Reiner, *Typo-graphik*, Verlag Zollikofer, St Gall, n.d.

Novices are mistaken when they suppose there should
be a "technical" term for every product of their enthu-
siasm & ignorance. Most terms of the craft are used &
illustrated in this book. Several are derivations from
the terminology of type engraving & the typefoundry.
"Ears' at the end of some curves (a, g), a "bracket"
on top of a t are familiar enough. Writing-
masters indulged in all kinds of curlicues which
were eliminated with the corresponding terms by
industrial thrift. We are left with quite a number
of strokes: horizontal, vertical & oblique, as well
as curves, bowls, tails, loops, spines & counterpunches. Which are not exotic terms & technical only
as far as they go,

Humanists
Garaldes
Transitionals
Didones
Slab-serifs

Lineales
Glyphics
Scripts
Graphics

The classification of typefaces is full of pitfalls. Especially for beginners who want to classify all typefaces in sight before they can even correctly identify one of them. No classification will ever satisfy all parties. The one Maximilien Vox (a French typographer, †1974) elaborated during the 1950s was internationally discussed in many professional papers before it was eventually adopted by A. Typ. I. Such as it is it invites perception of what is constantly there to be seen as well as read, & opens an historical perspective on an essential aspect of our daily environment,

The Italian humanists adapted to their purposes the 'Carolingian', a book letter used in ninth-century Tours to copy the Latin classics. Nicolas Jenson, Wendelin of Speyer, Ehrard Ratdolt & other printers in fifteenth-century Venice fashioned typefaces after these humanistic models. The resulting typefaces had a large x-height, robust serifs, little contrast. They were revived by William Morris, Emery Walker et al. at the end of the nineteenth century. Their characteristics are so marked that a humanist 'tendency' may even be attributed to the semi-bold versions of other families. Their capitals line with the ascenders (bdhkl) & therefore unduly catch the eye.

Contemporary humanists & their designers:

ALDUS, Herman Zapf 1954
ATHENEUM, A. Butti, 1941
AUGUSTEA, A. Novarese, 1948
CENTAUR, Bruce Rogers, 1929
DELPHIAN, R.H. Middleton, 1928
DE ROOS, S.H. De Roos, 1951
DIOTIMA, Gudrun Zapf, 1948
ELDORADO, W.A. Dwiggins, 1951
SISTINA, Hermann Zapf, 1950

EMERSON, Jo Blumenthal, 1935
ESTIENNE, G.W. Jones, 1930
EUSEBIUS, E. Detterer + R.H. Middleton, 192
GOUDY, F.W. Goudy, 1947
LUTETIA, Jan van Krimpen, 1930
DUTCH MEDIEVAL, S.H. De Roos, 1913
MERIDIEN, A. Frutiger, 1956
MICHELANGELO, H. Zapf, 1950
PALATINO, H. Zapf, 1951

ulla mentio erat . Quare nec iudæos(poft(
gentiles:quoniam non ut gentes pluralita
ìebræos proprie noïamus aut ab Hebere ι
:ranfitiuos fignificat.Soli qppe a creaturi
ìo fcripta ad cognitioné ueri dei trãf·ere:8
ìd rectam uitam pueniffe fcribunt·:cum q
totius generis origo Habraam numerãdus
ïuftitiã quã non a mofaica lege(feptima e:
Moyfes nafcitur)fed naturali fuit ratione
atteftatur.Credidit enim Habraam deo &
Quare multarum quoq; gentium patrem
ipfo benedicédas oés gentes hoc uidelic& ì
ìperte prædictum eft:cuius ille iuftitiæ pe
fed fide côfecutus eft:qui poft multas dei
filium:quem primum omnium diuino p
cæteris qui ab eo nafcerétur tradidit:uel a
eorum futuræ fignum:uel ut hoc quafi p:
tinétes maiores fuos imitari conaret·:aut q
enim id fcrutãdum nobis modo eft.Poft l
pietate fucceffit:fœlice hac hæreditate a p:
coniunctus quum geminos genuiffet caft
dicitur abftinuiffe.Ab ifto natus é Iacob (
prouétum Ifrael etiam appellatus eft duo

Roman used in 'De Preparatione Evangelica' by Eusebius.
Engraved and printed by Nicolas Jenson in Venice, 1470

HUMANISTS

The roman of Nicolas Jenson
(1419?-1480) inspired
William Morris, Emery
Walker & Cobden Sander-
son who led the typo/
graphic revival of the
late nineteenth century.
Note the so-called 'kerne
characters 'Qu' cut as
one 'sort' or character.
Also the 4 slab-serifs
of the M. The oblique
stroke in the 'e'. Also
the oblique slant of
the lower-case 'o'.

The garaldes belong to an Italo-French tradition which held sway during the sixteenth & seventeenth centuries, & regained popularity during the past 60 years. It originated in fifteenth-century Venice with Aldus Manutius & his engraver Francesco Griffo. It was rendered famous during the fifteenth century by Dutour, Garamond, Granjon, Le Bé; all of them worked for Plantin. Also by Caslon, Van Dyck & the Voskens during the eighteenth century. Compared with the humanists they look lighter & more contrasted. Their x-height is smaller. As a result there is also more contrast between the medials, the ascenders & descenders.

Contemporary garaldes & their designers:

ASTRÉE, R.Girard, 1928
BEMBO, Monotype, 1929
CRISTAL, Rémy Peignot, 1957
EMERGO, S.L.Hartz, 1947
I.T.C.GALLIARD, M.Carter, 1981
GARALDUS, A.Novarese, 1957
GRANJON, G.W. Jones, 1925

IMPERIAL, Ed.W.Shaar, 1954
IMPRINT, J.H.Mason + G.T.Meynell, 1912
ROMANÉE, J.van Krimpen, 1928
ROMULUS, J.van Krimpen, 1931
SPECTRUM, J.van Krimpen, 1942
TRUMP, Georg Trump, 1954
YENDÔME, François Ganeau, 1951

VYVANT & cor
courciſſement du fr
regardé directemen
faut que vous enten
tificiel, pourtant qu
ment, & autant qu'il
nt l'inconuenient qui en pourroit ac
nt(à fin que vous en donniez garde,
idrez faire en celuy Art) faut que voi
ir de vous, eſt enclos en vn cercle, d
t que la veüe ſe tourne, toutes choſes
iteſfois ne ſ'en peut rediger en port
ſe dudit cercle, dont vous meſmes [c
is faire mieux entendre ay merqué
trialle, laquelle vous ſeruira d'exemp
ier en l'erreur ſuſdit. Donc la platte fc
s par A.B.C.D: vous repreſentera coi
Colónes ou pilliers quarrez, merque:
q pilliers diſpertiz ſus l'vne des large
is faut maintenant trouuer le poinc

GARALDES

This roman was designed by or after Claude Garamond (1500?–1561). The type itself is lighter. The spacing is slightly wider. There are less contractions, abbreviations, ligatures, & logotypes. The distribution of the thicks & thins (or distribution of weight) is oblique.

an designed by or after Claude Garamond, used in 'Le Livre erspective' by Jean Cousin, Paris, Jean le Royer, 1560

The transitionals represent the middle of the eighteenth century. They originated in the late seventeenth century with the Romain du Roi. Commissioned by Louis XIV it was cut first by Grandjean, later by Jean Alexander & Louis Luce. This was also the period of Fournier, Baskerville, Fleischmann, Rosart, et al. The transitionals are narrower after the "goût hollandais" as it was called. The thicks & thins are more contrasted & have a vertical stress. They can still be said to reflect calligraphy. Especially the italics. Although very much Ancien Regime their clearcut contrast between thick & thin foreshadows the modern style. This is the largest family in current use.

Contemporary transitionals & their designers:

CORNELL, G.F. Trenholm, 1953
BASKERVILLE, Monotype, 1923
BELL, Monotype, 1931
DIETHELM, Walter Diethelm, 1955
ELECTRA, W.A. Dwiggins, 1935
FAIRFIELD, R. Ruzicka, 1948
GEORGIAN, G.W. Jones, 1932

JUBILEE, Linotype, 1954
MONTICELLO, Linotype, 1947
PAGANINI, Bertieri-Butti, 1928
PERPETUA, Eric Gill, 1934
PILGRIM, Eric Gill, 1934
TIMES NEW ROMAN, S. Morison 1932
WEIS ANTIQUA, E.R. Weiss, 1928

C. CORNELII

T A C I T I

H I S T O R I A R U M

L I B E R   P R I M U S.

I. Initium mihi operis Ser. Galba iterum, T. Vinius confules erunt. Nam, poft conditam urbem, octingentos & viginti prioris ævi annos multi auctores retulerunt : dum res populi Romani memorabantur, pari eloquentiâ ac libertate ; poftquam bellatum apud Actium, atque omnem poteftatem ad unum conferri pacis interfuit, magna illa ingenia ceffere.

A iij

*T*ACITI præfatio in Hiftorias fuas ; ejus dignitas, ætas, inftitutum. Præfentis Hiftoriæ fumma capita. Status urbis, mens exercituum, habitus provinciarum, occifo Nerone. Galba invifus ob fua amicorumque vitia ; inimicorum mores. Status Hifpaniæ, Galliæ, Germanicorum exercituum, aliarum provinciarum. Muciani virtutes & vitia. Vefpasianus in Judæa. Ægypti, Africæ, Italiæ &c. ftatus. Superioris Germaniæ miles deficit à Galba, qui de adoptando Cæfare cogitat ; magno amicorum ambitu, & fpe Othonis : fed Pifo eligitur, homo nobilis & feverus. Galbæ oratio, caufas adoptionis, monita imperii adminiftrandi continens.

Roman by S. P. Fournier the younger used in Tacitus, 'Historiae', Paris, J. Barbou, 1760

TRANSITIONALS

Abbreviations are totally abandoned. The spacing between the words exceeds the spacing between the lines (or leading) - a practice hardly to be recommended. 7 out of the 9 lines in roman, 10 out of the 15 lines in italic, end with a punctuation mark or hyphen. To-day only three consecutive divisions are tolerated & the space between the words should never exceed the width of an "e".

In the Romain du Roi Grandjean followed to the best of his ability a quite doctrinaire & outspoken geometric rationalism. The improvements of the printing press & even more so the smoothing out of the paper surface also smoothed the way for a wholly geometric tendency which is the characteristic of the didones. The initiator was F.A. Didot, in 1784. He was followed by Bodoni, Unger, Walbaum, Thorne, et al. They conquered Europe in the footsteps of Napoleon. In spite of their typefaces' fragility they had to endure the rough conditions of newspaper production. Serifs are reduced to a straight line. The stress of the thicks & thins is vertical; their contrast is violent.

Contemporary didones & their designers :

AMATI, Georg Trump, 1952
CALEDONIA, W.A. Dwiggins, 1938
COLUMBIA, W. H. Mc Kay, 1956
CHISEL, after R. Harling, 1939
EDEN, R.H. Middleton, 1934
EDITOR, Henri Chaix, 1937
NORMANDIA, Butti-Novarese, 1944

EGMONT, S.H. De Roos, 1933
IDEAL, E. Thiele, 1941
IMPRIMATUR, Bauer + Baum, 1952
OPAL, Weber, 1936
SAPHIR, H. Zapf, 1953
SMARAGD, Gudrun Zapf, 1954
VERDI, Bauer, 1957

DIDONES

Quousque tandem abutêre, Cati-
lina, patientiâ nostrâ? quamdiu
etiam furor iste tuus nos eludet?
quem ad finem sese effrenata ja-
ctabit audacia? nihilne te noctur-
num præsidium Palatii, nihil urbis vigiliæ, nihil timor populi, ni-
*hil concursus bonorum omnium,
nihil hic munitissimus habendi se-
natus locus, nihil horum ora vul-*

MARCUS TULLIUS CICERO

ORATOR ATQUE PHILOSOPHUS.

Even spacing, wide mar-
gins & borders meant
more to Giambattista
Bodoni than the sense
intended by an author.
Here again out of 10 lines
7 end with a punctuation
mark or a hyphen. But
spacing is even, while
there are only 6 words
per line, which is (with
all due respect) hard
enough.

*Roman from the 'Manuale Tipografico' of
Giambattista Bodoni, Parma 1818*

The slab-serifs are the easiest to identify as such. In the first place, of course, this is due to the serifs in question, but also to the typefaces' seemingly uniform strokes. In their typographic versions their origin is fairly recent they are contemporary with the Industrial Revolution (nineteenth century) & modern publicity. The romantics swooped on these serifs & gloriously used & abused them for the most exotic decorations. Unrivalled for sheer impact, slab-serifs are favourites with admen & circuses. Most newspaper typefaces are a blend of transitionals & slab-serifs.

Contemporary slab·serifs & their designers:

CENTURY, Benton-Delvinne, 1894
CLARENDON, H. Heidenbenz, 1952
CRAW, F.T. Craw, 1954
EGIZIO, Butti-Novarese, 1952
FORUM, G. Trump, 1948
SUPERBA, E. Thiele, 1932
IONIC, Linotype, 1925

MELIOR, H. Zapf, 1952
MEMPHIS, E.R. Weiss, 1929
PRIMER, R. Ruzicka, 1953
PRO ARTE, Max Meidinger, 1954
PROFIL, Eugen Lanz, 1947
QUIRINUS, A. Butti, 1938
SCHADOW-ANTIQUA, G. Trump, 1937

THO RNE shbo z urn

An Egyptian from Thorowgood's
'New Specimens of Printing
Types', London 1821

SLAB-SERIFS

There is nothing ambi / guous about slab-serifs. It does not follow that they are always big, bold, & neat. Some variant are light: the type-writer, characters for example. As such, slab-serifs have the widest circulation of all: if only through the mail. Slab serifs are also known as Egyptians.

The lineales make up the large & age-old family of the SANS i.e. without serifs, without contrast between thick & thin. Some architects in the wake of the Bauhaus (1919-1933), some painters in the wake of Dada & surrealism felt that lineales should be used as book-faces i.e. for continuous reading. To them serifs were superfluous ornaments incompatible with modernity. In this they were not followed by the bulk of publishers simply because serifs, thicks & thins are the very substances of typographic culture & readability as distinct from mere legibility.

Contemporary lineales & their designers

ANTIQUE ANNONCE, Amsterdam n.d.
ITC AVANT GARDE, H. Lubalin, 1970
KABEL, Rudolf Koch, 1927
CHAMBORD, R. Excoffon, 1945
FUTURA, Paul Renner, 1928
GILL, Eric Gill, 1928

NEUZEIT, W. Pischner, 1928
PEIGNOT, Cassandre, 1936
HELVETICA, Meidinger, 1957
UNIVERS, A. Frutiger, 1956
SCULPTURA, W. Diethelm, 1956

O BE SOLD WITHOUT RESERVE OUSEHOLD FURNITURE CLASS AND OTHER EFFECTS

LiNEALES

In the larger sizes lineales are main ly used for billboards & signposting. In book sizes they are not too popular with the literati. From what can be seen on the periodicals they seem to be more to the liking of scientists.

Sans-serif from 'Specimen of Printing Types',
by Vincent Figgins, London 1832

Glyphic (from the Greek γλυφη, gluphè, chiselled) evokes the incision of the chisel on both sides of the stroke & more specifically of the serifs. Long or short, the serifs are the essential feature in this family & are always pointed. Examples abound in epigraphy i.e. in inscriptions, old & new, on pediments, tombstones & copper engravings. In typography they are almost a rarity. Their typographic forefathers are to be found in the 'Latins' which were popular in the late nineteenth century. Glyphics do not invite continuous reading.

contemporary glyphics & their designers:

AUGUSTEA, A. Novarese, 1951
ALBERTUS, Berthold Wolpe, 1935
CAPITALES ECLAIREES, John Peters, 1957
COLUMNA, Max Caflisch, 1955
CRISTAL, Rémy Peignot, 1955
LARGO, Ludwig + Meyer, n.d.

LATIN ANTIQUE, Stephenson-Blake
MERIDIEN, A. Frutiger, 1957
PASCAL, José Mendoza, 1960
PRESIDENT, A. Frutiger, 1956
WEISS-LAPIDAIRE, E.R. Weiss, 1931
PHOEBUS, A. Frutiger, 1955

Monumentales simples et ombrées.

NOTE 89

Latines maigres.

LA REVUE
principal

Monumentales allongées.

L'INDUSTRIE

Latines étroites.

SAISON HIVER
L'Art d'Apollon

Monastiques simples et ombrées.

LÉON ATHOS

Latines larges.

A PARIS
en France

Latines noires.

CULTIVE
on jardin

Latines allongées noires.

ROME, FLORENCE

Latines blanches.

⁕ HENRI ⁕
PRIME

Latines éclairées.

HUILES D'IRIS

Latines maigres ombrées.

RHUM

Latines blanches à filets.

COLIQUE

Latines demi-blanches.

FRANCE

Latines ombrées.

OPÉRA COMÉDIE
le Théâtre ancien
DERNIÈRE
Nouvelle

Latines effilées.

COMPTE-RENDU OFFICIEL

*A variety of glyphics in F. Thibaudeau, 'La lettre d'imprimerie',
Paris 1921*

From these examples it
clear that glyphics are
part & parcel of the Fin
de siècle folklore of
French cheese & wine
labels. They are clearly
no more conducive to
continuous reading than
the marks of the chisel a
distinct from the pen
or burin. The various
shadings have a litho
graphic origin.

Scripts include all typefaces which simulate the kind of handwriting which used to be associated with the despatching of any & all official & commercial documents & correspondence. The first typographic version of such a script was the so-called CIVILITÉ of Robert Granjon, 1557. It was intended but failed as a book-letter. In 1643, Pierre Moreau cut a script which was closer to the documentary model. Scripts are not fit for continuous reading. Their typographic use is evocative & symbolic rather than functional. By now publicity is their province & they simulate the brush & the crayon rather than the pen.

Contemporary scripts & their designers:

CALLIGRAPHIQUE, Faulque, 1921
CHOC, Roger Excoffon, 1955
DELPHIN, G. Trump, 1950
DIANE, R. Excoffon, 1956
LEGENDE, E. Schneidler, 1937
MAXIM, P. Schneidler, 1956

MISTRAL, R. Excoffon, 1953
ONDINE, A. Frutiger, 1956
REINER, Imre Reiner, 1951
RAFFIA, H. Krijger, 1952
RONDO, Schlesinger + Dooijes, 1948
SLOGAN, A. Novarese, 1957

CARACCERE DE FINANCE,
Dit Bâtarde Coulée.

Nouvellement gravé par Fournier le jeune, Graveur et Fondeur de Caractere d'Imprimerie. Demeurant actuellement rüe S.t Etienne des grès, proche l'Abbaye de S.te Geneviève. à Paris. 1749.

Le present Caractere est gravé à l'usage des Imprimeurs curieux, pour l'impression de certains ouvrages legers qu'on voudroit faire passer pour être écrit. Il est utile pour les Epitres dedicatoires, Lettres circulaires, Billets de Commerce, d'invitation d'Assemblée, de Ceremonie, &c. Necessaire surtout pour les ouvrages d'Intendance, comme Mandements, Permissions, Ordonnances, Avertissements, Edits, deffenses &c. Pour les ouvrages du Secretariat des Evéchez, pour les Bureaux des Fermes, les Gabelles, les finances et autres que le goût & la Curiosité dicteront.

'La Bâtarde Coulée', cut by Fournier the younger, Paris 1742

scripts

Joseph Carstairs, 1783–1843? designed what he intended as a national (British) model for commercial & administrative use. It met with universal success. & is still known as 'copperplate.' In th[e] U.S., the Spencer & the Palmer methods are still in use for current handwriting as distinct from calligraphy

With or without serifs & in all their diversity, graphics reproduce the rhythm of formal writing. Before the advent of printing all book letters: uncials, carolingian gothics were 'graphics' as a matter of course. They had inevitably a personal touch in spite of considerable efforts to achieve the standardisation required for continuous reading. Modern typographic versions a. intended for publicity & titlings. They too simula. the brush & the crayon rather than the pen. A persona touch is deliberate. As a result they are no longer cor sidered suitable for continuous reading.

contemporary graphics & their designers :

BANCO, R. Excoffon, 1951
CODEX, G. Trump, 1953
CONTACT, I. Reiner, 1955
FLASH, Crous-Vidal, 1952
DOM CASUAL, Peter Dom, 1955
RITMO, A. Novarese, 1955
STUDIO, A. Overbeek, 1946

LASSO, M. R. Kaufman, 1939
JACNO, Marcel Jacno, 1954
LIBRA, S.H. De Roos, 1938
MERCURIUS, J. Reiner, 1957
PARIS, Crous-Vidal, 1952
RICCARDO, Richard Gerbig, 1928
PSITT, René Ponot, 1954

ene ✠ dic qñs sācte pater has cre
aturas herbarū. ut sint remedium
utate generi humano. et presta p inuo
ione sācti tui nomis. ut quicūq; ex eis
npserint. corporis sauitate; et anime
tutelam percipiant. Per dominū.
remus pietate tuā omnipotēs creet
deus. ut primicias creature tue quas
is et pluuie munere et temperamēto fru-
care et crescere iussisti. bene ✠ dictio-
tue ymbre perfundas. et fructus terre
; ad maturitatem pducas. tribuasq;
o tuo de tuis muneribus tibi sp gras
ere. ut a fertilitate terre esurientiū ani
s bonis affluētibꝰ repleas. ut egenus
aup laudēt nomē glie tue. p d. Et be
dictio dei pa ✠ tris. et fi ✠ lij. et spiritꝰ
sācti descēdat sup has creaturas amē.
inde thurificētur et aspergantur aqua
dicta.

y Johann Sensenschmidt, Bamberg 1491

Graphics

Gothic letterforms are
also called black letter
as against the roman
& italic which might
be described as blond
or fair. Textur, Fraktur,
Rund Gotisch or Schwa-
bacher, they all have a
marked rhythm too.
Their majuscules are
for use exclusively as
initials since they do
not constitute an inde-
pendent alphabet like
the roman capitals.

To summarise. All roman typefaces of the fifteenth, sixteenth, seventeenth & eighteenth centuries are fit to be used as body text & for titling. They can meet all reading situations: global, discontinuous, continuous. The SLAB-SERIFS, the LINEALES, the GLYPHICS are intended for titles, billboards & short paragraphs. The GRAPHICS, formerly bookhands, & SCRIPTS formerly cursives are now used for titling & publicity. ■ Typefaces are generally hybrids. So much is suggested by the rubrics GARALDES (Garamond + Aldus), DIDONES (Didot + Bodoni). Any model must always be adjusted to some purpose & technique. It takes as much time to develop a typeface, a book letter, as to write a novel. Maybe more. Depending on the letter & on the novel, of course. Type designers are clearly among the unsung heroes of contemporary art.

...nists, 15th–16th centuries *Garaldes, 16th–17th.* *Transitionals, 17th–18th.*

...dones 18th– th *Slab-serifs, 19th–20th* *Lineales, 19th–20th*

...lyphics, 19th–20th *Graphics* *Scripts*

This is "light"
but more
controlled than...

LIGHT

To choose a letter is not a problem. Eventual embarrasment is due to ignorance & the ensuing errors. This can be helped : by study, observation & experimentation. A bold face is for titling. Its place is on top. It also suggests weight. A light face suggests the opposite. Weight tends to sink. And vice versa. This kind of juggling belongs in the world of publicity. Some allusions may be even more far-fetched. And to edit Voltaire in gothic would be just as incongruous as to compose the Bible in a light slab-serif.

;, which is hurried
not controlled
all.

Light

e material elements of any transcription should
er be interpreted as hindrances but as helpful compo-
nts of any solution. Strict sobriety in the choice of
aterials is likely to lead to an elegant solution. To
ake indiscriminate use of all available resources
characteristic of beginners: this blurs the message
istead of clarifying it. Letterforms are all that are
quired to enliven a reading surface. Give them
ace. Let them sparkle. One type for titles, one for the
t will suffice to articulate quite a few messages.

Read:
Sobriety.

Titles & texts should be ranged along one or two main axes. No more. Lest the whole printed surface gets mix ed up. Some even avoid indenting paragraphs to preserve the integrity of justification. For centuries this had been the whole truth as far as text pages & even title pages were concerned. Then by the end of the nine- teenth century a painter, Whistler, & a poet, Mallarmé, shattered the age-old symmetry. (Before 1914, Francis Thibaudeau invented the asymmetric 'typographie des groupes'. In the 1930s, Jan Tschichold (1902-1974) was the official exponent of asymmetric typogra- phy (later he recanted & died as a devotee of tradi- tion & symmetry). Yet even Tschichold was content with 2 axes & a composition fitted into the bottom half of the page. However, he made sure to have a contrast in almost every line of a title page → bold-roman, roman-italic, caps-lowercase. And out of four lines, three are flush left.

Real size:
280 x 200 mm

The Penrose Annual

Review of the Graphic Arts   Edited by R. B. Fishenden, M.Sc (Tech) FR PS

Volume Forty | 1938

_LUND HUMPRIES & Co LTD   12 Bedford Square, London, W1

Tracing-paper was introduced into typographic usage about 1900 by Francis Thibaudeau. It met with the contemporary needs of asymmetric typography & proved so useful generally that it soon spread all over Europe & the U.S. The whole typographic environment has changed beyond recognition. But even to-day simulation is necessary before work commences, Tracing paper is also a good tool for the self-taught. There is no lack of it. There are also plenty of pencils, crayons & chalk. Printed models are all over the place. You take your choice. Our example is simple: an horizontal ad 2 col. wide to be resized 1 col. wide. A first sketch to number & place the lines along a central axis. Next it is written out. Notice that 'Le présent avis' is even more legible & eye-catching in one line of italics than in the original 2 lines of uppercase ■ It is not always as simple as that. Basically one tracing is required for every single component, These are subsequently assembled together so that everything fits into

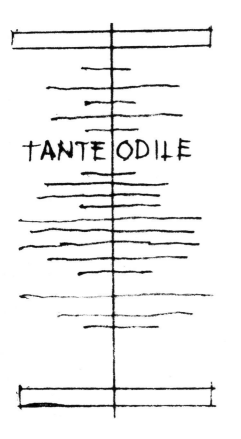

place. We have seen from the example of a daily newspaper
how crucial tracing-paper layouts are for processing
& assembling the copy, for keeping track of proofs at
various stages, for marking in the codes & instructions
at each stage of production.

Francis Thibaudeau, *Manuel français de typographie moderne*, Paris, 1924.

Type ANTIQUES, *ou* LETTRE BATON, *sans empattement.* (Tracé Phénicien.)

INE ine

Type ÉGYPTIENNES. *Empattement quadrangulaire brut.* (Tracé Grec.)

INE ine

Type ÉGYPTIENNES ANGLAISES. *Empattement quadrangulaire et arrondis d'angles intérieurs.*

INE ine

Type ELZÉVIR. *Empattement triangulaire.* (Tracé de la capitale romaine d'inscriptions.)

INE ine

Type DIDOT. *Trait fin horizontal d'empattement.* (Caractéristique : opposition de graisse des pleins et des déliés.)

INE ine

Type HELLÉNIQUES. *Empattement triangulaire et jambages bi-concaves.* (Tracé au calame.)

INE ine

Type TRAIT DE PLUME, *à caractéristique elzévirienne d'empattement triangulaire.*

INE ine

The A.Typ.I. classification is not the only one in existence. It was not the first. It is not the latest. Thibaudeau was the first to attempt a rational classification of typefaces. His one criterion was the serif. Which is a bit on the short side. By way of compensation it looks simple & convincing if not exhaustive. The A.Typ.I. classification takes into account any number of criteria, historical as well as visual. It is not simple. Twenty years later, Jean Alessandrini, a man of letters in both senses, put forward a new classification, CODEX by name (see: COMMUNICATION & LANGAGES, n° 43 + 44). It introduces additional criteria, subjective as well as geographic. Since then manufacturers & suppliers have proposed a variety of classifications better suited to their own & their customers' needs. To be sure, to place an order it is safer to refer to the retailer's reference whatever that is rather than to any particular classification.

Why do we have such a mass of typefaces? sometimes ⫫
identical even though under different names. Because
typefaces are a commodity in the same way as motor
cars & furniture. Because there is no way to reduce
all alphabets to only one style of upper & lower case
any more than to reduce all furniture to a single
model of a table or chair. Moreover free trade favours

a spirit of competition, of compromise or of fraud depending on temperament, & circumstances. As a
result today each typeface has a proper name regard
less of its size, its designer, its owner. It may be re-
tailed by any number of licensees. It can be pirated
& sold under any number of spurious names. One
aim of A. Typ. I. is to seek legal protection internationally. Another is to develop a critical knowledge
of typography among the public. In all this the
rational rubrics of a classification system help describe the particulars of typefaces in general terms & pro

inéales   French

gothics   American English

sans   English

groteske   German

·de a basis for discussions. The rubrics of the Vox
ε·A.Typ.I. classification have been officially trans-
ated into 7 languages. Even so because of the
omplexity of its cultural references it has no place
· daily practice. It cannot be enforced. It is not
·clusive of other classifications. Yet it still makes
·nse as a re-introduction to literacy in the full sense.

The traditional sym-
metrical layout of a
double spread. When
there are 2 or 3 col. to
the page, the 4 or 6 col.
of the double spread
constitute ONE visual
unit.

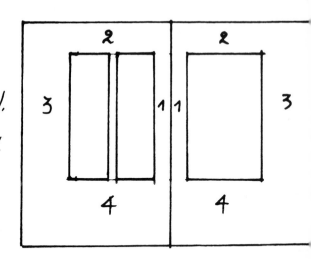

News governs the news desk & the newspaper. This results
a daily helter-skelter... & the absence of margins. The ulti-
mate raison d'être of the book format is study. Hence, sobriety is the rule. In 1924, Marius Audin said: "Serious
printers lavish the utmost care on imposition which is
a key factor in an elegant book." He took 6 pages and
4 figures to make his point. W. Morris made his point
in the following words without any illustration: "The
inner margin should be the narrowest, the top somewhat wider, the outside (fore edge) wider still, & the

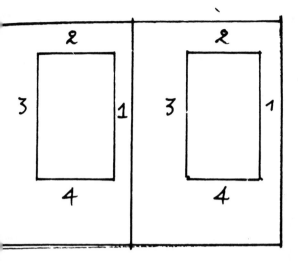

A modern, asymmetrical layout. To some people any asymmetrical layout looks modern. This is not the case: asymmetry & modernity do not work like magic formulas.

ottom widest of all. This rule is never departed from in medieval books, written or printed. Modern printers systematically transgress against it; thus apparently contradicting the fact that the unit of a book is not one page but a pair of pages." (Intelligenti pauca). Before 1920, there was no rationale of asymmetry. Since then there has been no lack of theories, styles, rational & irrational, symmetric & asymmetric. The difference is not a matter of difficulty. It is a matter f typographic culture... & literacy tout court.

End-matter

How this book was written produced.

In the original French edition this was an assignment. The subject matter was drawn from ten years' experience as a teacher on the blackboard in front of 18-25 year old students in product design. No one expected me to instruct would-be typographers or graphic designers. Product designers had to be made type conscious as typographic communicators & aware of the fact that printed matter is an industrial product in its own right. The school * had no print shop worth mentioning. That is how & why the blackboard was my main resource (also because it happened to be there as in any class room or laboratory.

François Richaudeau, the French publisher, wanted a handwritten book. I hesitated because I am not a trained calligrapher. We compromised: part of the book was typeset, part of it handwritten. In the handwritten section a loose sequence of double spreads was the obvious

- Hooger Architektuurinstituut van het Rijk (HAR) Afdeling Produkt-ontwikkeling, Antwerpen.

choice. First because the double spread is the visual unit of a book anyway. Second because this is not a sequential story. Still less a history of its subject. Each double spread makes its point about a particular aspect of a vast subject. By way of crude sketches such as anyone could make on a blackboard. Also by way of printed matter. Just as I illustrated my teaching with printed examples: finished products of all kinds & origins. Also my own work in progress & in its various stages: typewritten copy, tracing paper & layouts, proofs, etc. All this as an introduction to the practical work of the students.

The printed section.

In this we followed current practice. First a manuscript, then a typed copy duly marked up & cast off; followed by proofs & corrections

The handwritten section.

This was a bit more complicated. As I have said I am not a trained calligrapher. Still less a professional one. This was my first commission for a handwritten job. It shows clearly in all details! Anyhow I had to take the plunge. A first draft was written out & illustrated, just to SEE what happened, how it worked & HOW IT LOOKED. This was submitted to the publisher. And eventually approved without corrections. Then came the real thing. In order to introduce the required visual coherence, I naturally applied the age-old (timeless indeed) practice of the grid. Not a ready-made, standard formula—such a thing does not even exist—but the ad hoc thing. So, I cut paper to the final size of the printed book & traced the text area & its margins. The size of the writing, the length of the lines & the space between the lines ('leading') were established by trial & error. I pasted the resulting grid on to my desk following my familiar writing-angle. The leaves for the copy were in turns placed over the grid. Some pages had to be written over several times before they came out right. All in all I found the challenge more enjoyable than crosswords (which I do not enjoy at all).

All this is a bit crude—including the result. But apparently it works. The transcription took a dozen days of sustained work. Writing & designing the book was an altogether different matter. It took on & off six months maybe? The English version was much simpler, since apart from problems of translation & substitution of English examples for those in French, I had the French version as a model.

the illustrations.

They are of two kinds. Those which are reproduced with permission & those made as best I could. The first question was: should all my sketches be reproduced out of a black background to simulate chalk on a blackboard? The answer was emphatically NO. In the first place reading a book is a totally different experience from reading a blackboard. See reproducing & photographing one by one all the lessons as they were given would have been exceedingly slow & expensive. Therefore I made my approximations with pen & ink on paper & as crudely as anyone could make them on a blackboard.

Select bibliography

English-language titles

Aldis, Harry G.
The printed book
Cambridge University Press 1951

Bartram, Alan
The English lettering tradition
Lund Humphries, London 1986

Beaujon, Paul [Beatrice Ward]
'The "Garamond" types: sixteenth and
seventeenth century sources
considered'
The fleuron, V

Berry, W. Turner, A. F. Johnson &
W. P. Jaspert
An encyclopaedia of typefaces
Blandford Press, London 1970

Bigelow, Charles et al. (ed.)
*Fine print on type: the best of Fine Print
magazine on type and typography*
Bedford Press, San Francisco/
Lund Humphries, London 1988

Bland, David
The illustration of books
Faber & Faber, London 1962

Bland, David
*A history of book illustration: the
illuminated manuscript and the printed
book*
Faber & Faber, London 1969

Brown, Bruce
*Browns index to photocomposition
typography*
Greenwood, Minehead 1983

Butcher, Judith
Copy editing: the Cambridge handbook
Cambridge University Press 1982

Carter, Harry
*A view of early typography up to about
1600*
The Lyell lectures: Clarendon Press,
Oxford 1968

Carter, Sebastian
Twentieth-century type-designers
Trefoil, London 1987

*The Chicago manual of style for
authors, editors and writers*
University of Chicago Press
(13th rev. ed.) 1982

Child, Heather (ed.)
*Formal penmanship and other papers:
Edward Johnston*
Lund Humphries, London 1971

Child, Heather & Justin Howes
*Lessons in Formal Writing:
Edward Johnston*
Lund Humphries, London 1986

Craig, James
Production for the graphic designer
Watson-Cuptill, New York 1974

Crutchley, Brooke
*Preparation of manuscripts and
correction of proofs*
Cambridge University Press 1967

Day, Kenneth (ed.)
*Book typography 1815–1965 in Europe
and the United States of America*
Ernest Benn, London 1966

De Vinne, Theodore Low
*Modern methods of book composition:
a treatise on type-setting by hand and by
machine and on the proper arrangement
and imposition of pages*
New York 1904

Dreyfus, John (ed.)
*Type specimen facsimiles: reproductions
of fifteen type-specimen sheets issued
between the sixteenth and eighteenth
centuries*
London 1963

Dreyfus, John (ed.)
*Type specimen facsimiles II:
reproductions of fifteen type-specimen
sheets issued between the sixteenth and
eighteenth centuries*
London 1972

Fowler, H. W.
A dictionary of modern English usage
Oxford University Press (rev. ed.) 1965

Garland, Ken
*Graphics, design and printing terms: an
international dictionary*
Lund Humphries, London 1988

Gill, Eric
An essay on typography
Lund Humphries, London (rev. ed.)
1988

Glaister, Geoffrey
Glaister's glossary of the book
Allen & Unwin, London 1979

Gowers, Sir Ernest
The complete plain words
Penguin Books, Harmondsworth
(rev. ed.) 1973

Hart, Horace
*Hart's Rules for compositors and
readers at the University Press, Oxford*
Oxford University Press, Oxford
(rev. ed.) 1983

Hostettler, R.
The printer's terms
SGM, St Gallen 1959

Hutt, Allen & Bob James
Newspaper design today
Lund Humphries, London 1988

Jennett, Sean
The making of books
Faber & Faber, London 1973

Johnson, A. F.
*Type designs: their history and
development*
Andre Deutsch, London 1966

Kindersley, David & Lida Lopes
Cardozo
Letter slate cut
Lund Humphries, London 1981

Lowry, Martin
*The world of Aldus Manutius: business
and scholarship in Renaissance Venice*
Basil Blackwell, Oxford 1979

McLean, Ruari
Jan Tschichold: Typographer
Lund Humphries, London 1975

McLean, Ruari
*Modern book design from William
Morris to the present day*
Faber & Faber, London 1958

McLean, Ruari
*The Thames & Hudson manual of
Typography*
Thames & Hudson, London 1980

Miles, John
Design for desktop publishing
Gordon Fraser, London 1987

Morison, Stanley
*On type designs past and present: a brief
introduction*
Ernest Benn, London 1962

Morison, Stanley
First principles of typography
Cambridge University Press (2nd ed.)
1967

Morison, Stanley
A tally of types
Cambridge University Press 1973

Perfect, Christopher (ed.)
Rookledge's International Typefinder
Sarema Press, London 1983

Ryder, John
Printing for pleasure
Bodley Head, London (rev. ed.) 1976

Simon, Oliver
Introduction to typography
Faber & Faber, London (2nd ed.) 1963

Spencer, Herbert
The liberated page: a Typographica anthology
Lund Humphries, London 1987

Spencer, Herbert
Pioneers of modern typography
Lund Humphries, London (rev. ed.) 1982

Spencer, Herbert
The visible word
Lund Humphries, London (rev. ed.) 1969

Steinberg, S. H.
Five hundred years of printing
Penguin Books, Harmondsworth (rev. ed.) 1974

Sutton, James & Alan Bartram
An atlas of typeforms
Lund Humphries, London 1968

Swann, Cal
Techniques of typography
Lund Humphries, London (rev. ed.) 1980

Tracy, Walter
Letters of Credit: a view of type design
Gordon Fraser, London 1986

Tschichold, Jan
Assymetric typography
Faber & Faber, London 1967

Twyman, Michael
Printing 1770–1970: an illustrated history of its development and uses in England
Eyre & Spottiswoode, London 1970

Updike, Daniel Berkeley
Printing types, their history, forms, and use: a study in survivals
Dover, New York/Constable, London (rev. ed.) 1980

Williamson, Hugh
Methods of book design
Yale University Press, New Haven and London (rev. ed.) 1983

French-language publications

Baudin, Fernand
La lettre d'imprimerie
Ets Plantin, Brussels, 1965

Baudin, Fernand & John Dreyfus
Dossier mise en pages
Magermans, Andenne 1972

Diethelm, Walter
Emblème, Signal, Symbole
Editions ABC, Zürich 1970

Dreyfus, John & François Richaudeau
La chose imprimée
Retz, Paris 1977

Fertel, Dominique
La science pratique de l'imprimerie
Fertel, Amiens 1723

Fournier Le Jeune, Pierre-Simon
Manuel Typographique utile aux gens de lettres, 1764

Frutiger, Adrian
Type, Sign, Symbol
Editions ABC, Zürich 1980

Gerstner, Karl & Markus Kutter
Le nouvel art graphique
Niggli, Teufen 1959

Gorce, Maxime
Les pré-écritures et l'évolution des civilisations
Klinsieck, Paris 1974

Haab, Armin & Walter Haetten Schweiler
Lettera 1, 2, 3 et 4
Niggli, Teufen 1961

Havelock, Eric A.
Aux origines de la civilisation écrite en Occident
Maspéro, Paris 1981

Hoffmann, Armin
Manuel de création graphique
Niggli, Teufen 1961

Imprimerie Nationale
De plomb, d'encre et de lumière
Paris 1982

Jayan, Alfred & Guy Delage
Ecriture et structure
Petite bibliothèque Payot, Paris 1981

Morison, Stanley
L'art de l'imprimeur: 250 reproductions des plus beaux spécimens de la typographie depuis 1500 jusqu'à 1900
Dorbon-Aîné, Paris 1925

Morison, Stanley
Les premiers principes de la typographie
Translated by F. Baudin in: *Stanley Morison et la tradition typographique*
Catalogue d'exposition, Bibliothèque royale Albert Ier, Brussels 1966

Morison, Stanley
Caractères de l'écriture dans la typographie
A l'enseigne du Pégase, Paris 1927

Ruder, Emile
Typographie
Niggli, Teufen 1967

L'Ecriture et la psychologie des peuples
Centre international de synthèse
Armand Colin, Paris 1963

Thibaudeau, Francis
La Lettre d'imprimerie
2 vol. Paris 1921
— *Manuel français de typographie moderne*
Paris 1924

Vox, Maximilien
Faisons le point
Union Bibliophile de France, Paris 1983

Glossary

Alignment. To line up type or illustrations on common vertical or horizontal.

Alphanumeric set. Complete sequence of alphabet (upper- and lower-case) plus matching numerals, accents and punctuation marks.

Ampersand. Symbol for 'and'; &.

Antique paper. Uncoated paper with matt finish, often quite bulky, used typically for books without illustration.

Art paper. Coated paper with glossy surface, particularly suitable for illustrations.

Ascender. Part of lower-case letters extending above the x-height (in letters b, d, f, h, k, l, t).

A Typ I. *Association Typographique Internationale,* founded in 1957 'to bring about a better understanding of typography, a higher level of typographic design, and to secure international protection for typefaces'.

Black letter. Type based on fifteenth-century style of calligraphy, also known as 'graphic'.

Bleed. To allow image to run up to edge of page.

Body size. Measurement in points of vertical dimension of typeface.

Body type. Type used for main text, as distinct from display matter.

Bold. Heavier version of roman type.

By-line. Line of type giving name of author of newspaper article.

Calligraphy. Art of fine handwriting.

Capital. Upper-case letter, A, B, C etc (also known as majuscule).

Caption. Description accompanying illustration.

Carolingian. Script developed in Europe during the reign of Charlemagne.

Casting-off. Estimating how much space a text will occupy when set in a certain size, to a certain measure and a certain number of lines per page.

Chancery italic. Style of handwriting of 15- 16th centuries, on which italic typefaces were based.

Character. Individual letter, numeral or punctuation mark in a particular typeface.

Coated paper. Paper having a surface covering made from china clay, suitable for illustrations.

Colophon. Publisher's device used as a 'trademark' on his publications.

Colour (of typeface). Appearance of typeface in terms of lightness/heaviness.

Composition. Act of assembling type characters, by hand or machine.

Computerised composition. Assembling type characters via a computer keyboard.

Condensed. Narrow version of typeface.

Copy. Matter to be typeset.

Counter. Space enclosed by bowls of characters a, b, d, e, g, o, p, q.

Cpi. Characters per inch.

Cuneiform. Wedge-shaped form of characters inscribed on ancient clay tablets.

Cursive. Typeface which simulates handwriting, particularly one based on chancery italic.

Descender. Part of lower-case letters appearing below the x-height (in letters g, j, p, q, y).

Desk-top publishing. Term covering office systems whereby text (and possibly illustrations) are generated by computer and made up into finished pages for reproduction.

Didones. Style of typeface dating from late 18th century, with fine hairlines and unbracketed serifs (also known as modern face).

Display matter/size. Text, usually above 14pt in size, used for headings.

Dropped capital. Initial capital of paragraph set in larger size of type, extending over several lines of body size.

Dummy. Book or publication produced in advance as mock-up to indicate specification and appearance.

Egyptian. Slab-serif typefaces, originating in 19th century.

Elite. Smaller of two standard typewriter alphabet sizes.

Em. Dimension derived from square of a given type size; 12pt em (known as pica em) is used to measure length of lines and areas of type.

En. Half the width of an em; average width of character, and used therefore in typesetting calculations.

End-matter. Items such as bibliography, index, etc which appear at end of book after main text.

End papers. Sheets at each end of casebound book which fasten pages to cover.

Exotic typeface. Term applied by European printers to typefaces other than those using Latin letterforms, e.g. Arabic, Hindi.

Expanded/extended. Wider version of typeface.

Face. Short for 'typeface'.

Family. Complete range of characters available in a particular typeface, including roman, bold, italic, upper-case and lower-case.

Fat face. Typeface in which contrast between thick and thin strokes is particularly marked.

Figure. (1) numeral (2) line illustration.

Fixed word spacing. Method of typesetting in which spaces between words are all standard, giving rise to unjustified (or ragged right) lines of text.

Flush left/right. Lines of type extending to common vertical, i.e. justified.

Folio. Page number.

Font (also fount). Set of characters from typeface in one size, incorporating upper-case, lower-case, small capitals, numerals, punctuation marks and possibly other signs too.

Foot. Bottom of book; hence 'foot margin' is the margin at the foot of the page.

Foredge. Edge or margin of book opposite the spine.

Format. Traditionally, dimensions of page or book. Also, with development of computerised typesetting, now refers to particular constellation of characters or configuration of text determined by computer program.

Front matter. Also referred to as prelims, short for 'preliminary pages' which precede main text and consist of items such as half-title, title, list of contents, etc.

Frontispiece. Illustration in prelims, usually facing title page.

Full point. Full stop.

Galley proof. Proof·of text in form of long strip, produced before text is made up into pages.

Garalde. Family of typefaces also known as 'old face' or 'old style'; typical is Garamond.

G/m². Abbreviation for grams per square metre, measure of weight of paper in UK and Europe.

Gothic. Another term for 'black letter' except in the USA where gothic refers to sans-serif typefaces.

Graphic. Typeface such as 'black letter' which reproduces the rhythm of formal writing; used nowadays for display purposes, not for continuous reading.

Grid. Unifying system governing arrangement of text and illustrations on the page in a particular publication.

Grotesque. One of family of sans-serif typefaces.

Gutter. Inside margin of book nearest to spine.

Half-title. The first right-hand page in book, usually containing title only, set small.

Hanging indent. First line of paragraph overhanging rest of text in left-hand margin; the rest of the lines appear indented.

Head. Top of book, opposite foot.

Hot metal. Type cast in metal, as distinct from typeset photographically or by computer.

House style. System of rules governing grammatical usage and appearance of type in books and other publications.

Humanists. Italian Renaissance and particularly Venetian designs, dating from the 15th century and based on a revival carolingian text.

Hyphenation. Method of breaking and joining words, using hyphens; particularly important for word breaks at end of lines.

Imposition. Arrangement of pages on printing sheets so that when sheet is folded pages will come together in correct sequence.

Imprint. Publisher's or printer's name printed on book.

Incunabulum (*pl.* incunabula). Early examples of printing, especially applied to books printed before 1501.

Indent. To leave blank space at start of line of type, typically first line of paragraph.

Italic. Typeface originally based on chancery italic handwriting though now applied to typefaces with characters which incline to right; most typefaces appear in roman and italic forms.

Jobbing printer. Non-specialist printer who takes on variety of small printing jobs.

Justify. Drive out lines of type to common measure, achieved by varying space between words.

Kerning. Adjusting space between individual letters to achieve closer fit.

Keyboarding. Selecting and assembling characters in passages of text, using a keyboard on a typesetting machine or computer.

Laid paper. Uncoated paper which reveals faint pattern of lines when held up to light.

Layout. Plan of publication produced by designer as working guide for printer, indicating how printed results are to be achieved.

Leading. Originally metal strips inserted between lines of type to create space; still used to refer to space between lines, measured in points.

Leaf. Single sheet of paper consisting of two pages which back up.

Legibility. Ease with which text can be read.

Letterpress. Printing process which involves printing from a raised surface.

Letter-spacing. Spaces inserted between letters within a word.

Ligature. Two or three characters joined together as one to avoid awkward spacing between letters, e.g. fi, ff, ffi.

Lineales. Sans-serif typefaces.

Lining figures. Numerals all the same height which line up with capitals.

Lithography. Printing by the planographic process in which parts of the printing plate accept ink and other parts reject it.

Majuscules. Upper-case characters, otherwise known as capitals.

Make up. To arrange type and illustrations in pages.

Manuscript. Written matter submitted for typesetting.

Mark up. Supply typesetting instructions on manuscript or typescript.

Matrix. Mould from which metal type is cast; master image for photographic composition.

Measure. Length of line of type, usually expressed in pica ems (12pt).

Minuscules. Lower-case characters.

Newsprint. Paper used for newspapers.

Old-style figures. Numerals which unlike lining figures do not all range with capitals but have ascenders and descenders; also known as non-lining figures.

Opacity. Quality in paper which prevents ink or images showing through from one side of leaf to the other.

Optical spacing. Space introduced between letters to take account of their different shapes and to give appearance of even spacing.

Page. One side of leaf of book.

Paginate. To number pages of book.

Papyrus. Writing surface made from reeds in ancient Egypt.

Paste-up. Text and illustrations pasted in position on pages either as guide for printer or for actual reproduction.

Photocomposition. Typesetting by photographic means.

Pica. (1) unit of typographic measurement [12pts] (2) larger of two sizes of typewriter face.

Point. Basic unit of typographic measurement. There are approximately 72 points in one inch.

Prelims. Short for preliminary pages, which precede main text.

Proof. Version of text or illustrations produced by typesetter or printer for checking purposes before printing commences.

Ranged left/right. Text lined up on common vertical, at start or end of line.

Ream. Unit used in ordering paper (approximately 500 sheets).

Recto. Right-hand page of book (carries odd page numbers).

Relief printing. Printing process (e.g. letterpress) in which raised surface is inked and printed.

Reverse out. Type which prints white out of a coloured background.

Roman. Normal, as distinct from bold or italic, characters.

Running headline. Title of book or chapter repeated at head of page for guidance.

Sans-serif. A typeface without serifs (strokes at ends of letters).

Script. Typefaces which imitate handwritten forms.

Serif. Stroke at ends of type character.

Set. (1) to typeset (2) width of a character.

Sheet. Paper used for printing, containing a multiple of pages (e.g. 8, 12, 16, 20, 32).

Slab-serif. Typeface with square-ended serifs.

Small caps. Capitals which approximate in height to x-height, unlike full-size capitals which correspond to x-height plus ascenders.

Solus position. Advertisement which appears on a newspaper page where there are no other advertisements.

Sort. Individual letter, numeral or punctuation mark in a particular point size.

Spread. Two facing pages, left and right.

Stet. Latin for 'let it stand': instruction to ignore proof correction previously marked.

Swash letter. Ornamental character, usually italic capital, with emphatic flourish.

Tail margin. Bottom margin of book.

Text type. Typeface and size used for main text, usually exceeding 14pt.

Title page. Usually second right-hand page in book, carrying details of title, author and publisher.

Transitional. Family of typefaces having origins in 18th century (e.g. Baskerville), falling between garaldes (or old-style) and didones (or modern).

Type area. Area on page containing text matter.

Type family. Variants of a particular type design, encompassing (if they all exist) light, medium, bold, italic, condensed and expanded versions.

Type mark-up. Instructions for typesetter marked on typescript or manuscript.

Type scale. Rule or gauge used by typographers and calibrated in points and ems as well as inches and milimetres.

Type specimen. Sample produced by manufacturers of type design to show complete alphanumeric range.

Typographer. Graphic designer who designs piece of printed matter (but in USA, a compositor; one who sets type).

Typography. Study and practice of design and use of typefaces.

Unjustified. Text with ragged right-hand edges to lines (i.e. with even word spacing).

Upper-case. Capital letters.

U & l.c. Mixture of capitals and lower-case letters.

Venetian. Typefaces originating in 15th-century Renaissance Italy, particularly Venice, known as humanists.

Verso. Left-hand page of book (carries even page numbers).

Weight. Referring to visual appearance of typeface, whether heavy or light.

Wf. Abbreviation for wrong font, proof correction mark indicating that letter from another type family has strayed into a piece of setting.

Word processor. Computer which can be used to generate, update, correct, rearrange, store and print out text.

Wove paper. Plain paper without laid lines.

x-height. Height of lower-case letters without ascenders or descenders.